AF690689

Darwin's Incomplete Idea

Wittgenstein, language, our place in nature and our responsibility for the environment

Gunnar Odhner

VERNON PRESS

Copyright © 2015 Vernon Press, an imprint of Vernon Art and Science Inc, on behalf of the author.

All rights reserved. No part of this publication may be reproduced, stored in a retrieval system, or transmitted in any form or by any means, electronic, mechanical, photocopying, recording, or otherwise, without the prior permission of Vernon Art and Science Inc.

www.vernonpress.com

In the Americas:
Vernon Press
1000 N West Street,
Suite 1200, Wilmington,
Delaware 19801
United States

In the rest of the world
Vernon Press
C/Sancti Espiritu 17,
Malaga, 29006
Spain

Library of Congress Control Number: 2014948701

ISBN: 978-1-62273-143-5

Product and company names mentioned in this work are the trademarks of their respective owners. While every care has been taken in preparing this work, neither the author nor Vernon Art and Science Inc. may be held responsible for any loss or damage caused or alleged to be caused directly or indirectly by the information contained in it.

Table of Contents

Chapter 1
An increasing concern

In the mid-1980s two concerned persons meet in Stockholm and Moscow to discuss the growing threats to the environment. They are the Swedish author Rolf Edberg and the Russian ecologist Alexej Jablokow. Both are deeply worried about the deterioration of nature.

Their conversations are recorded in a book titled *Sunday is too late.* Nine tenths of the text can be regarded as a catalogue of all the threats to our environment a generation ago. The remaining part is an attempt to explain why these threats occur in the first place.

Why bring up a book published 30 years ago? Because it gives us a perspective. What has happened with the environment in a generation? What has been done to counteract pollution and the deterioration of natural systems? Are we moving or standing still?

The first thing the Swedish and Russian ecologists discuss is the danger of nuclear warfare. They ask why the arms race is still on, even though everybody knows that a war with nuclear weapons will be devastating. East and West are locked in antiquated thinking, they say. And so is the rest of the world. During 1986, a designated "year of peace", 36 wars and armed conflicts were fought. Five million soldiers were involved.

Today the nuclear problem still exists. There are more countries today that have nuclear weapons than in 1986, among them unstable states like Pakistan and North Korea. The international community also worries about the nuclear capability of Iran. Even so, the threat of nuclear warfare seems less today than in 1986 when the world was locked in the cold war and feared nuclear annihilation. The new nuclear powers are not yet capable of causing global destruction. In the area of armed conflicts and wars the situation is roughly the same. According to UCDP, Uppsala Conflict Data Program, the number in 2013 was 33.

In 2015 wars are going on in many parts of the world, especially in Syria, Nigeria, Yemen, Sudan and in the eastern part of Ukraine. 10 million people have fled their homes in Syria. New terror organisations like the IS have popped up alongside the old ones. The conflict between Israel and

Palestine (the roots of which can be traced to the late 19th century!) is still unsolved. Contrary to what is accounted for in popular books conflicts and violence seem to be steadily increasing. 51.2 million people were "forcibly displaced" worldwide in 2013, the highest figure since World War II. (UNHCR).

Nuclear energy is another problem according to the two ecologists. They argue that it has to be replaced by renewable sources and point to Harrisburg and Chernobyl to strengthen their argument. If they had lived now they would certainly have added Fukushima.

The population growth is another threat. They quote a UN-report that predicts that 12 to 14 billion people will inhabit the earth around 2050. "This cannot go on", Edberg says, and emphasizes that a life-style like his own on a global scale would ruin our planet.

These figures have been revised lately. According to Hans Rosling, the Swedish professor of international public health, the correct assumption is around 9 billion. In this area of world development we have good reasons for optimism and hope for the future!

The greenhouse effect and climate change are next on their list. If the two environmentalists had been contemporary, they would have talked even more about it. Today this problem is of the utmost concern. The Swedish EPA says that it is desirable that emissions of greenhouse gases should peak before 2020 and then be halved before 2050. This does not seem likely to happen. The goal of 2 degrees centigrade 2020 is in danger.

China alone has increased its emissions of greenhouse gases from 14 per cent of the world's total emissions in 2000 to 27.3 per cent in 2013. The US emits around 18 per cent.

Another concern that is addressed in *Sunday is too late* is the exploitation of the rain forests. "Every year rain forest the size of Austria disappears." (RE).

When I write this Austria is replaced by the slightly smaller Panama. If this exploitation continues there will be no rainforests at all a hundred years from now.

The increasing use of fresh water is another item on the list. "The majority of the industrial countries use 500 litres per day and person" (AJ). This is unattainable worldwide.

For different reasons many species of our flora and fauna are threatened. "At the beginning of the twentieth century it was estimated that one species a year was extinct. Now in 1985 this figure is one every day. The earth is like a sinking arc."(RE).

This problem has grown worse. Every 20 minutes a species becomes extinct today. The contemporary widespread poaching in Africa is a serious threat to elephants and rhinos. There were once over 100,000 black rhinos on the plains of Africa. There are now less than 3,000 on the entire continent. The staggering decimation of the rhino population is largely due to poaching, to satisfy the demand for the horn for use in Eastern traditional medicines.

Between five to ten million African elephants existed in 1930. More than 100 000 were killed between 2010 and 2012 (Save the Elephants, Aspinall Foundation). China destroys ivory, but the trade, according to these organisations, is out of control. Today most countries in West Africa count their elephants in tens or hundreds. Probably only three countries in this region have more than 1,000 animals. In contrast, elephant populations in southern Africa are large and expanding, with some 300,000 elephants roaming across the sub-region (WWF).

Recent figures from the WWF show that the number of tigers is increasing. 510 tigers live in Siberia. India, Nepal, Bhutan and Bangladesh have 2 800 tigers together.

On the Internet there is a list of close to 11,000 seriously threatened species.

In the depressing list of the two ecologists one part is devoted to pesticides and all kinds of litter that is harmful to the environment, plastic for example. The amount of plastic in the Pacific has an area as large as Texas. Artificial fertilizers is another problem, especially for lakes and seas such as the Baltic. The long list of grievances ends with a discussion of the ozone-layer and the situation of the poor in large parts of the world.

The environmental problems that Edberg and Jablokow worried about in 1986 (Edberg already in 1966) are almost the same today or have, in some respects, worsened.

In Sweden the Environmental Protection Agency points out that 14 of the nation's environmental goals will not be reached in 2020. That means two fulfilments.

And Sweden reputedly is in the forefront of environmental protection!

The time to solve the ecological problems is running out, Edberg and Jablokow say. Edberg said the same thing in 1966 in his book *The Remains of a Cloud.* "We must be in peace with the forces of nature . . . the task is enormous. Maybe it has to be solved during the next generation." Nearly half a century later Johan Rockström and Mattias Klum, wellknown environmentalists of today, say the same thing: time is running out. But there is one big difference. We do not have the time we had in the 1960s or 80s. "We must turn the tide of negative change before the end of this decade . . . establishing global environmental goals that the world must adhere to in order to avoid catastrophic system changes." (JR, MK 2012).

What was impending in 1966 and 1986 has become even more so. Why did not the process of mending the weakening systems start 50 years ago? Or with the first serious warning signals? These warnings were not figments of the imagination. Many capable voices called out as early as in the 1960s. In many aspects of environmental protection time seems to have been standing still. According to Rockström and Klum, human interference goes on undiminished. It is difficult to avoid the conclusion that something is in the way of environmental work, a hindrance that the persons involved are not aware of.

Some people think that this is so because "human beings don't change". I think this is the worst conclusion we can make as it stands in the way of progress. This book is devoted to showing the opposite: that humankind is almost infinitely malleable. The reasons for the tardiness of environmental work cannot be sought in a "human nature". They are far more complex and have to do with the construction of language/consciousness and the structure of our societies, which I hope will become obvious in the following chapters.

What then are the causes of the environmental damages according to Edberg and Jablokow as well as most ecologists? Mainly two: ignorance and hubris.

"Limited knowledge and bottomless ignorance are equally responsible for our predicament." (RE). "It is the lack of wisdom in using our knowledge that has created our dilemma. We do not understood what we are doing." (RE).

The same thoughts are repeated by modern ecologists. "Humanity has placed itself in a planetary crisis due to ignorance, lack of understanding, unwillingness and passivity." (JR, MK). It is not until now that we have the knowledge needed to wholly understand the ecology of nature, these ecologists say. If this is so, environmental work should take a giant leap ahead in our time. The complete picture is there! What are we waiting for?

Can we trust this explanation of environmental deterioration? Is it true?

It seems to me as if it contains a blatant willingness to put the blame on others. *They* were the ones who did it, not we. It was *the others,* the ignorant and arrogant ones. *They* lacked the knowledge and wisdom that we now have. *We* try to mend what *they* have broken.

If this explanation is true, environmental work should have picked up momentum from the moment the enlightened wisdom was evident. It does not seem to be the case. On the contrary. From the ominous development of the threats I have mentioned it almost seems to be the other way around: the more knowledge the slower the environmental efforts.

A terrible thought strikes me, the more so as I am influenced by Socrates: Could it be that this knowledge is not *real* knowledge, but rather something bearing that name?

If it was real knowledge, would it not compel us to decisive action?

Now it is more like the knowledge that the Egyptian king Thamus warned the god Theuth of in Plato's dialogue *Phaedrus.* Theuth had invented writing and came with his invention to Thamus. He said that his discovery would "improve both the wisdom and the memory of the Egyptians." Thamus turns this into the opposite. "Those who acquire writing will cease to exercise their memory and become forgetful." They will stop using "their own internal resources." He continues: "And as for wisdom, your pupils will have the reputation for it without the reality: they will receive a quantity of information without proper instruction . . . and because they are filled with the conceit of wisdom instead of real wisdom they will be a burden to society." Is this applicable to the environmentalists of today? It seems to me that it might be the case. Even so, the words of Thamus are certainly relevant today with the domination by media and the ease of consulting the internet. Could it be that we have built a huge temple of knowledge that is empty behind the walls, the way Thamus fears?

In his discussion of knowledge, following the metaphor of the cave, Plato makes a distinction between what might be called "superficial knowledge" and "innate knowledge", where the former is obscuring the latter. In order to have real or innate knowledge, Plato says, the whole body must participate: the heads of those who can see only the shadows on the wall, not their origin, have to be turned. This cannot be done with compulsion. "Compulsory physical exercise does no harm to the body, but compulsory learning never sticks in the mind." Knowledge has to be integrated with the body, buried in our hearts and visible in our actions in order to be knowledge! "The capacity by which a man learns is like an eye which cannot be turned from darkness to light unless the whole body is turned", Plato concludes.

This idea of knowledge, of all of us *being* knowledge rather than just having it, was later developed by Sören Kierkegaard in his exploration of subjectivity. And it is also visible in the philosophy of Ludwig Wittgenstein, who seems to say that we *are* language. Maybe we need more of this perspective in order to succeed in our relation with nature and each other?

After all, if we look at the objective knowledge of our cosmos, in what way does it compel us to act? Scientist agree that the universe is cold and indifferent. Why then should we care about it? Where is the connection between a callous universe and caring people?

Let me rephrase the question: Do we really have the full knowledge to act, which is what ecologists contend? Do we know our position in nature? Do we really know the origin of ourselves? How life started? What life is? Do we even know what a human being is?

There is another flaw in the explanation made by the ecologists: the talk of human imperfection, of arrogance and unwillingness as part of human nature.

This is, as I see it, an appalling approach. If it comes down to fixed human characteristics and human nature we will never solve our problems, because passivity will go on. You can say, with a twist of the argument, that it is nature's own fault that it suffers.

This viewpoint is common among biologists and evolutionary psychologists, at least among those I have read. To them it is self-evident that human beings have a nature. How could we have risen from nature if we do not? They see human nature as playing a vital part in the making of history and many of them agree with Richard Dawkins in *The Selfish Gene*: "unfortunately we were born selfish." As I see it this is a feeble attempt to explain why human beings behave contrary to their own interests. It is not a correct view which I will show later on in this book. Again: if we were born destructive we are not to blame, but nature is.

There have been other attempts to explain why environmental work is sluggish. One such theory concerns the Prisoner's Dilemma, a game theory developed in the 1950s. According to the theory individual countries may deem it advantageous to betray the collective, at least for a shorter period of time, thus obstructing a development that is beneficial to all. The theory is based on what the creators call "rational self-interest" of the prisoners. But as far as I can see they never explain what they mean by "self-interest", never show from where it stems and what makes it exist in the first place. It is rather the base of their idea as if it was an immutable part of human nature. I will later show that the idea of a human nature is very dubious. If that is the case it undermines the theoretical foundation of the Prisoner's Dilemma. Self-interest as I see it is caused by the way our societies are structured – it is not self-interest that structures our societies. The creators of the PD seem to regard it as primary. In my opinion it is secondary, a social product of vital importance in all unequal societies.

If there is such a thing as rationality in connection with self-interest it is best described by Baruch Spinoza in his Ethics 350 years ago. It seems as if self-interest is dissolved in the process, that it cannot co-exist with rationality no more than can infatuation.

> Therefore, to man there is nothing more useful than man – nothing, I repeat, more excellent for preserving their being can be wished for by men, than that all should so in all points agree, that the minds and bodies of all should form, as it were, one single mind and one single body, and that all should, with one consent, as far as they are able, endeavour to preserve their being, and all with one consent seek what is useful to them all. Hence, men who are governed by reason – that is, who seek what is useful to them in accordance with reason, - desire for themselves nothing, which they do not also desire for the rest of mankind, and, consequently, are just, faithful, and honourable in their conduct.

It seems as if our environmental problems could be easily solved, because as many ecologists put it "we have the knowledge" and "we are in the driver's seat". If we take the right steps in the near future our environmental problems will soon be solved. This is, I think, to overlook the complications, the greatest one being the knowledge of our own position.

We are surely in charge, but in what way? I think that our relation to ourselves and our surroundings is much more complicated than the ecologists believe: we are not only in control, we are also controlled. How this mechanism works I will explain later.

But if this is so, we are not only authoritative, acting subjects. We are also victims. And as such we are probably entitled to more compassion and love than blame and guilt.

What is the nature of the knowledge that we now at last have, the knowledge that eventually will save us? It can be summed up in two words: ecology and ecosophy.

"Ecology is the science of the interactions between the living creatures and their surroundings. Ecology shows us how every species is related to other living creatures and to earth, air and water." (RE in *A House in Cosmos*).

Ecosophy on the other hand is simply ecological philosophy. It can be explained as "the wisdom about the house." (RE). Ecosophy is characterized by an urge to "reach a survey, a total picture". It is an expression of a "cosmic-biological totality that encompasses everything from atoms to galaxies." (RE). Notice that the author talks about all living creatures, including humankind. It means, if we take him seriously, that ecology shows us the truth about our relationship with other creatures and earth, air and water. Does it? Or is the ecological idea more like a promise or a hope, a first step on a road that will lead to clarity?

We are introduced to a picture of the world that encompasses everything. How is this possible, when large parts of the universe are still unknown and new theories of the birth of cosmos are presented almost every day? Is total knowledge even possible?

Edberg envisions the world we live in as a house that we know from the basement to the attic. This knowledge is necessary in order for us to skilfully manage "the house".

Exactly the same thought is put forward by other ecologists nearly 50 years later. Rockström and Klum ask for a "global durability agenda that moves us away from the exploitation of natural resources" for the benefit of durability strategies. As managers of the planet, they say, we must "pay full attention to the resilience of the planet."

We have moved a long way from the local, simple attachment to soil, air and water into global perspectives and caretaking. Now it almost seems as if we must have our heads among the clouds or even the stars to be able to understand ourselves and our surroundings.

We must know our "house" in its planetary totality to be able to manage it wisely.

This totality of knowledge seems to be more attractive than the Star of Africa diamond. Many ecologists have made painstaking efforts to understand and map our "house" in its entirety, among them Rolf Edberg, a meticulous architect. He really tried to build the cosmic overall view that he, like many others, deemed necessary for a life in peace with nature. In the following chapters we will join Edberg and others to see if their knowledge is viable enough to enable a wise administration of our "house in cosmos".

Chapter 2
The cornerstone of life

"Life was born out of chemical processes, maybe in the meeting-place between sea and land and atmospheric gases, where ultraviolet sunlight and intense electrical charges decomposed inorganic matter into a brew, the parts of which were rearranged. The first signs of life must have been a vague impulse to react on the environment." (RE)

According to the text, this cosmic breakthrough took place around three billion years ago. Contemporary figures are closer to 4 billion years.

> The cosmic laboratory starts to produce proteins and other organic substances that occupy the shallow water with nutritious matter. It is not yet life but 'life at the threshold of life'. The first step is taken on the road that will lead to people, poets and dictators. We roughly know that it must have happened in this way. (RE)

The picture we get is that of a factory that blindly starts to produce goods of a kind called life. "We roughly know that it must have happened in this way." Do we? What does it mean to know "roughly"? You travel a slippery road when you know "roughly".

In an attempt to find a foundation for his house, Edberg calls life "part of the structure of the universe". If so the first signs of life must be the consequence of this structure. Just as we look at ourselves as products of the evolution, life is a product of the universe.

This is not a new idea. Darwin himself imagined that life started in this way, even though he was reluctant to dwell on the thought and regarded it as speculative.

In a letter of 1871 he writes that life probably originated in a warm pond, where a brew of phosphoric salts, light, heat and electricity existed for a very long time until the first elements of life came together in more complex molecules. This idea came to dominate the following century and still dominates in our time. But the question of the origin of life was of no great importance to Darwin. He brushed it aside in a few sentences. "At the present it is meaningless to brood on the origin of life. You might as well ask for the origin of matter."

So far no one has presented us with the first elements of life. The scientists pursue the answer along different roads. Some think that RNA was first and DNA second. A scientist in England managed to produce parts of synthesized RNA in 2009. Like so many others that try to create life in the laboratories, he thinks that it is merely a matter of time before such attempts are successful. There will soon be a reliable theory of the origin of life.

Richard Dawkins supports the RNA-theory in his writings, because the DNA-molecule needs enzymes to replicate itself. Proteins can handle this, but they need DNA to do it, so it is a catch 22 situation. Dawkins also thinks that chemists soon will be able to fully reconstruct the events that led to the development of the natural selection of plants and animals.

But we still do not know how life originated. Maybe we will never know. Is it possible to objectively answer the question: "what is life?" We *are* life ourselves, as it seems always in the middle of it. To be able to answer the question we must answer yet another, equally difficult: "what are we?" When we recognize ourselves as being life an objective answer is hardly possible. It is as difficult as examining the ground under our feet without moving. The question "what is life?" looks for an answer such as "this is life". Is such an answer possible? It seems as if the question "what is life" only can be answered by living the answer.

Those who try to explain go in the opposite direction, objectifying the target of their investigations. Life is seen as a separate entity, something that can be isolated from the living creatures. This makes the concept more abstract and thus more difficult to understand: as if to explain life you first have to get rid of it. The quotations from Edberg and many others all emphasize the perspective: "life is structure", "life is a system of chemical processes", "life is a coherent stream that flows through the ages". (RE).

Unfortunately Darwin himself at least once lost his footing in the same way. In *The Origin of Species* he writes about the overwhelming beauty of life, "which was originally blown into a few shapes or in one only". He separates life from its organisms and looks upon it as independent, something that was "blown into" the creatures from somewhere else. Dawkins, who discusses this passage of Darwin's book in one of his own is convinced that it is a concession to religion, which seems to be true, especially as Darwin was regretful.

In a letter to his friend Joseph Hooker a few years after the publication of his famous book, Darwin states these regrets. He is sorry that he used the words from Genesis when he really meant that life had arisen out of a completely unknown process.

I think the Austrian philosopher Ludwig Wittgenstein can help us understand why it is so difficult to define life. He often talks of us being led astray by language and that we are bewitched by it. Is this what is happening here? Because we have the word "life" and use it in many different contexts, we draw the conclusion that it must stand for something tangible, in the same way as the word "table" stands for something we can see and touch. But defining "life", making it more tangible, might lead us away from an understanding rather than closer to it. Wittgenstein showed this with many examples. We know well what is meant by the word "game", but when it comes to finding a reliable definition problems mount, the demarcations get blurred or too many, and what we sought for slips away.

But Wittgenstein was unknown to Edberg, so he will not give up. Something called life is there. It must have a definition! Life is something "that bubbles up from sea and earth in individuals, who gleam a moment and then disappear." He looks upon these individuals, including human beings, as "temporary vehicles of life" (which reminds us of Dawkins idea in *The Selfish Gene*, where he regards human beings as temporary vehicles of genes).

Edberg stresses the importance of integration. He tells us that the biosphere is "a single organism in which individuals and species play the same role as cells and organs in the individual's body . . . to be human means to live in intimate relationship with everything on this cosmic splinter." I can only add: if this is what it means to be human, nature would need no rescue operations. The biosphere would be like a self-playing piano!

Most scientists agree with Edberg. But there are exceptions. The Swedish physicist Ulf Danielsson was asked "what is life?" during a seminar. He ducked like Darwin saying that he would rather not deal with such questions. But others do. In a book with the title "*What is Life, in Cosmos, in the Cell, in Human Beings*?" published by Karolinska Institutet, one of the leading medical universities in the world, the authors reflect on the lack of a definition. "It is quite remarkable that a uniform and generally accepted definition of life does not exist". Life to them is a biological mechanism. But will such a reduced definition suffice?

In the 1940s the physicist Erwin Schrödinger also tried to answer the question "what is life?" His purpose was to show that the "life processes were no more mysterious than the principles that make a windmill, an alarm clock or a combustion engine work."

If he is right it seems very strange that no one so far has been able to come up with a definition. It is not very difficult to define a mill or a combustion engine.

Two other scientists, Lynn Margulis and Dorion Sagan, tried the same thing 50 years later without success. It is my guess that the search will go on for a long time with the same result.

So, others have tried what Edberg was doing. But he was more persistent than most of them. He simply would not give up. In his hunt for a reliable, final definition life is given enormous proportions. It becomes what I would call a "primary subject", towering above the "secondary subjects" of human beings. Like God or Nature with a capital letter.

"In this gigantic diversity life itself is primary, its forms secondary." (RE). If we are secondary how are we to take the full responsibility for ourselves and our surroundings? That responsibility would then ultimately rest with something else, with "life".

The question "what is life?" seems to lead us away from life and living human beings, transforming us into products of some undefined entity called life, which will be, in a matter of time, described on the pages of a chemistry book. Will we have our answers then?

Edberg and all who seek the answer to the question "what is life?" think so. They go on to burying human beings and their subjectivity in the "gigantic interactions of life". No place is left for us where we can feel at home, really experience what it means to belong. "In the seething eternity of life and in the vast connections between all forms of life we can no longer be self-centred. *There is nothing to refuse and nothing to desire.*" (RE, my italics).

These words speak for themselves. What is finally achieved is the very opposite of what is sought for. The overall picture of the house leads to the annihilation of life, of the persons inhabiting it! The abilities that make us human are gone: the power to refuse or desire.

Another author might put us on the right track. Vladimir Nabokov has said: "What you can control is never fully real; what is real you can never fully control."

Or define. Or maybe not even understand in a scientific way?

Every time I read a book by an ecologist or biologist, the first ones as early as in the 1970s, I repeatedly wrote in the margin S.K., the initials of Sören Kierkegaard.

I saw their texts and their view of the world through the eyes of the Danish philosopher, thereby discovering what he had seen in Hegel: a full-blooded speculator, a person who has lost his footing on the ground and thus becomes an easy target of the irony in one of Kierkegaard's masterpieces: the two volumes of *Concluding Unscientific Postscript.*

Speculators are persons who have cut the ties to their own existence, Kierkegaard argues, and who therefore live speculatively, lacking an earnest rapport with themselves. Some of their favourite expressions are "us", "we", "man", "the human race" and "the human species" which all give an impression of omnipotence, but also of being part of a fellowship.

"Such is life", and "people are like that" are typical expressions that the speculative persons use to remedy their own ignorance and give the impression of knowledge. Expressions such as "from a cosmic point of view" or "from a biological standpoint" seem to make speculators spokespersons of the whole of creation. Edberg writes in one of his early books: "From a cosmic point of view life has no beginning and can have no end."

Commenting on Hegel, Kierkegaard quietly inserts that only human beings have points of view. People put together speculative ideas as if behind their own backs, he says. They then pretend that they had nothing to do with it: cosmos or nature gave them the ideas.

A full-blown speculator tries to objectively speculate his way to answers and aspects without involving himself. He leaves out the intimacy, the embracement of the problem, the will to make it an integrated part of his own actions. He does not live his knowledge.

In *Concluding Unscientific Postscript* Kierkegaard describes how the speculator transgresses borders in such a way that he will, in the end, make himself ridiculous.

> If a dancer can jump very high we admire him, but should he, even if he could jump higher than any other dancer, give the impression that he could fly – then let laughter bring him down. To jump means essentially to belong to the earth and respect gravitation, making the jump momentous, but to fly means to be separated from earthly conditions, which is reserved for winged creatures and maybe the inhabitants of the moon – who knows, maybe the system will find their true readers there. One has done away with being human, and every speculator confuses himself with humanity, becoming something enormous and at the same time nothing. In his distraction he mixes himself up with humanity in the same way as the opposition papers say "we" and the skipper "bloody me". But . . . when you have learned that every schoolboy can say "we", you will understand that it means a little more to be *one,* and when you see that every bar-owner can personify humankind, you finally realize that it is more to be a simple human being than to play party games in that way.

The speculator firmly believes in his own ability to fly and does not seem to have the means to reveal what he is really doing, therefore being an expert in devious shifts of meaning, giving speculative words and ideas an aura of knowledge. What at first is called "hunch" or "probability" suddenly becomes knowledge and truth.

> Science has not only given us a clearer view of the evolution that leads to ourselves and the interactions of nature. It has also opened the vast cosmic expanses, slightly opened the door to the workshop of nature and has let us sense something of the elements that make up everything, energy and matter, organisms and non-organisms. (RE).

The speculator persuades himself to believe that he knows more than he really does. Like some scientists he pretends that there is a direct link between him and nature, making his findings untainted by his own participation, the concepts and structure of his language.

A speculative person is thus nowhere to be seen even though he is everywhere.

An ecologist like Rolf Edberg (and so many others) can be seen as a true speculator when he builds his "house in cosmos" with imaginative bricks. In a roundabout way he is himself aware of his position. The callousness with which the "house" is described might nourish moods of depression, he fears. If the individual sees himself as a "temporary habitat for the seed of life" on an insignificant planet bound for its death, as is the universe, he might believe that life has no meaning and that it makes no difference how he acts. However, the remedy is near at hand. It comes with the cause of the depression. If we look at the world and ourselves in a gloomy way, we will miss the "core of the evolutionary aspect". We can find the meaning of life in "the unceasing, through billions of years ongoing process on this planet". In these eulogies, small masterpieces of poetry, Edberg anticipates the writings of Richard Dawkins.

Being part of this breath-taking process should make us "thrill with reverence and humility". Out of nowhere feelings of awe rise to enrapture us. It seems as if Edberg is disbanding his own assertion that humans have nothing to desire or deny.

Human beings, trying to understand their place in nature, have arrived at something that can be called secure knowledge, Edberg maintains: "After many attempts to find his place in nature man has through the theory of evolution gained a firm ground to stand on."

Consequently we need no longer have doubts about our identity. "The human body is the product of a long evolution; it can be no different with our nature." (RE).

Our next task then is to scrutinize the being who in due time will abandon ignorance and arrogance to become a wise manager of the "cosmic house".

Chapter 3
Homo sapiens moves in

Even though the "house in cosmos" has an unstable foundation in the unanswered question "what is life", there are other rooms. One is the birth of humankind.

> In a world where the struggle for life during millions of years has created a balance between organisms from the virus to the whale, from the shell to the eagle, a swaying creature suddenly and by chance rises from the green shadows in the forest of time and starts searching for food in a constantly changing environment. An upstart in creation conquers the globe, multiplies and replenishes the earth. (RE).

There is a dubious turn in this picture of the introduction of humankind, something that is left unexplained. Although insignificant, this detail is still important. The balance is there and has been for millions of years. Now it is up-ended by an upstart "who conquers the globe".

Confronted with this description we should ask ourselves: How is this possible?

The upstart evidently belongs to nature, but is also, in Edberg's view, outside to a certain extent. The newcomer is not be controlled like other members of the balance.

Edberg's account of "primitive man" is paralleled by many other authors on the subject. On the one hand we are told that humankind rose from nature and was in harmony with it. We are connected to everything else in the creation. This connection is very concrete, because it is part of our bodies. "Humankind like all other mammals is genetically programmed for a natural condition with clean air, with varied colours in the landscape, above all green in different shades, and with all the noise and movement of animals around, especially that of the birds." (RE). Not only do we live in the green paradise. We are an integrated, inseparable part of it. "The primitive man felt concord with the creating forces of nature." (RE).

We are told that this is still so: under the layer of civilisation we are still nature's children. There are "secret bonds between nature and the processes of our souls." (RE). The roots are deep and impossible to remove: "To be true to the earth is to be true to one's own nature. From the time of our lives with nature its rhythm is innate in us." (RE).

This is the overture to our story: we are integrated parts of the great natural balance. But there is a continuation, quite different from the initial heavenly part. It runs parallel to the first, like two themes in music intertwining and at the same time independent. But in this second part of the story humankind is regarded as something outside nature, aliens detached from their natural habitat, longing for their roots. "We must once more find the concord with the creating forces of nature, the ones that the primitive man instinctively felt." (RE)

The alienation is even more ominous. The human being, who only a few pages ago was unquestionably integrated, is now something that is not needed at all. Humankind, Edberg says, can be "more easily spared in the totality than the worm". Our extinction is no tragedy.

We belong but are dispensable is the conclusion. Edberg has even likened us to "a cancer spreading on the face of the earth". With those words he torpedoes his own definition of ecology. It would show, he said, "how every species relates to other living creatures and to earth, air and water." Now it seems to show the opposite; how unrelated we are.

And he is far from alone to voice that idea. Many ecologists, biologists and scientists are of the same opinion. As early as 1864 Georg Perkins Marsh saw humankind as a nuisance. "Of all organic creatures man is the only one who can be regarded as a mainly destructive force." Georg Henrik von Wright, the successor of Wittgenstein at the University of Cambridge, regards humankind as temporary and is convinced that our annihilation "in the cosmic perspective" is no more important than "a pipe of snuff". Other authors agree. "For the sake of the earth humankind is not needed." (Reidar Ekner), "Humankind is an illness of the earth." (Kay Curry-Lindahl). "All that humankind finds out sooner or later must have a negative effect, deteriorate nature . . . If humankind disappears . . . it is an insignificant event that has no influence whatsoever on the world as a whole." (Bengt Hubendick).

Some of these ecologists turn the screw even more. First they tell us, like Edberg, that humankind is an integrated part of nature. Then they point to all the destruction caused by us in nature and conclude that it is not really our fault: nature made us destructive.

"The cities of humankind, the technological advances, the pollution and the horrible wars – all this is a product of our brain and thus a product of nature itself." (Stefan Edman, well-known Swedish ecologist). How is it possible that nature can initiate its own destruction?

Not only are we unnecessary, but our existence is a direct threat to the nature that we are indissolubly part of. The balanced nature seems to up-end its own balance.

How did this unbalance come about? Edberg, just like many other writers, points to a climate-change that in some mysterious way should have pushed humankind out of the ecological balance. Why then are we pushed out without being pushed out?

It seems that Edberg's writing on this point, like that of so many others, is an intellectual shipwreck. He first shows humankind "in concord with everything". At the same time we are aggressive aliens. "It was a beast of prey, an eater of carcasses and a killer ape that left the shadowy existence of the jungle." (RE). From the start weapons were in our hands.

A more modern Swedish author, the anthropologist Lasse Berg, is opposed to this view. He says that humankind was peaceful from the beginning, "made for a minimum of work and a peaceful life in small, constantly food-gathering flocks." He first came upon this thought during an interview with Richard Leaky in Nairobi. Leaky convinced him that the human being never was a killer ape. But Berg and Leaky essentially have the same problem as Edberg. It remains for them to show how this peaceful being could be a warmonger.

The thoughts of Edberg, Berg and Leaky are inspired by Darwin as are similar headstrong attempts to concurrently assert our concord with nature and our alienation.

In *The Decent of Man*, 1871, Darwin tries to pinpoint the nature of humankind. In doing so he sees empathy as an innate and dominant trait that evolution has brought about. He argues much like modern anthropologists: Humankind is predominantly social, sociable and empathic because of our development in peaceful flocks. But like Edberg and many others Darwin does not adhere exclusively to this standpoint. When he starts talking about the relationship between different races this innate empathy is nowhere to be seen. Human beings, who at first seemed to be one race, are suddenly divided into different groups hostile to each other. Darwin predicts that the more civilized races almost surely will annihilate the "wild" ones. What happened to the innate empathy that everyone is supposed to be endowed with?

In my opinion it is no longer Darwin the scientist who writes this but Darwin the member of a colonial power, a person who turns his back on empathy as if it never existed. "When civilized nations come into contact with barbarians the struggle is short, except where a deadly climate aids the native race", we can read in *The Descent of Man.*

Edberg and Darwin show us two incompatible pictures of humankind. In one of them we are in harmony with the environment, embedded in it. In the other the situation is reversed. Biology plays an important part in the construction of these pictures. We have, through Mother Nature, been endowed with "obscure tendencies that explain the hostility of groups and nations, the acts of war and our willingness to let security rest on terror and guns." (RE).

I think this is a cynical, callous and superficial opinion. Edberg blames nature and regards the "obscure tendencies" it has given us as the main cause of violence and bloodshed.

We are of course involved and have a certain responsibility but we are not ultimately responsible, according to Edberg and many others. Nature is behind it all. If we are the products of nature in the way he an so many other ecologists and biologists think, we must draw the conclusion that we are only partly responsible.

Darwin and Edberg share the same problem. In *The Origin of Species* Darwin depicts the rock foundation of the selective process: "Natural selection will never produce in a being anything injurious to itself, for natural selection acts solely by and for the good of each."

According to this "law" of nature humankind would not exist!

If Darwin's "law" of natural selection is true, then genocide is impossible. If war is possible, Darwin's "law" is false and we are not products of the evolution. But Darwin soon forgets that humankind cannot injure itself and, like modern ecologists, is torn between two opposing views. On the one hand we are part of nature, one the other we are exiled. In *The Descent of Man* he even looks condescendingly at humankind in a way that resembles the words of Edberg, von Wright and others. We are one of nature's failures.

> For my own part I would as soon be descended from that heroic little monkey, who braved his dreaded enemy in order to save the life of his keeper, or from that old baboon, who descending from the mountains, carried away in triumph his young comrade from a crowd of astonished dogs – as from a savage who delights to torture his enemies, offers up bloody sacrifices, practises infanticide without remorse, treats his wives like slaves, knows no decency, and is haunted by the grossest superstitions.

It is evident that Darwin shows contempt for "the savage", even if he, according to his own principles, should not, because the savage is also a product of evolution.

We meet the same contempt in the writings of his disciples, and even more explicit – a contempt turned into something much uglier. In two of his books, *The Splinter of a Cloud* and *At the Foot of the Tree*, Edberg embraces race biology. With the creation of humankind nature has gone astray and it is now our duty to find its right course. "We are forced to, when it comes to the limitations of our own species, to take destiny in our own hands." We have to do this because we have destroyed "the barriers that came with natural selection".

In the future, he speculates, we cannot rule out that there will be "a qualitative selection to stop the deterioration of our race by a random proliferation of our gene pool".

These words are not meant as a warning but rather as a recommendation. "When natural selection no longer rejects the defective and less viable, you cannot rule out that man himself, organized in a welfare society will put himself in the place of natural laws."

Can you imagine a welfare society that kills "the defective" and "less viable"? A society that decides which offspring will live and which will not? As if Edberg does not understand what he is discussing, he goes on repeating the same thoughts.

> It is now widely recognized that the medical and hygienic progress that has softened the harshness of natural selection, has led to a deterioration of our race that by the next millennium might have a snowballing effect. It might be regarded as necessary to turn the development in the right direction.

Had he never heard of racial institutes or selections made only 30 years earlier by Hitler's regime, aimed to stop a "widely recognized race deterioration"?

These ideas stem from a long and soiled tradition that originated with Darwin.

In *The Descent of Man* he differentiates between civilized and wild races and predicts, as we have seen, that the former will eradicate the latter. He does this although he sometimes seems to say that all of humankind is one race. The right thing would be, he writes somewhere, to divide the human race into subspecies. This tendency, to divide and select, is noticeable even when it comes to social issues. Racial differences in people play a part in the development of society. The conclusion in *The Descent of Man* is that racially deficient people should not be allowed to have offspring for the sake of mankind's welfare.

> Both sexes ought to refrain from marriage if they are in any marked degree inferior in body or mind; but such hopes are Utopian and will never be even partially realized until the laws of inheritance are thoroughly known…
>
> The advancement of the welfare of mankind is a most intricate problem: all ought to refrain from marriage who cannot avoid abject poverty for their children; for poverty is not only a great evil, but tends to its own increase by leading to recklessness in marriage. On the other hand, as Mr Galton has remarked, if the prudent avoid marriage, whilst the reckless marry, the inferior members tend to supplant the better members of society. Man, like every other animal, has no doubt advanced to his present high condition through a struggle for existence consequent on his rapid multiplication; and if he is to advance still higher, it is to be feared that he must remain subject to a severe struggle. Otherwise he would sink into indolence, and the more gifted men would not be more successful in the battle of life than the less gifted. Hence our natural rate of increase, though leading to many and obvious evils, must not be greatly diminished by any means. There should be open competition for all men; and the most able should not be prevented by laws or customs from succeeding best and rearing the largest number of offspring.

These are the first signs of social biology, later developed by powerful Darwinians such as Sir Julian Huxley, who advocated prohibition of marriage of the unfit, segregation of institutions containing degenerate individuals and sterilization of the unfit, proposals to be appalled by. He became the first director-general of UNESCO and argued that "a considerable percentage of the population is not capable of profiting from higher education."

Darwin seems to be trapped in his own evolutionary device. Little does it help that he mitigates his own words by saying, at the end of the quoted passage, that when it comes to the higher part of humankind's nature "other agencies are more important". By this he means "habit, reasoning powers, instruction, religion, etc." But he still maintains that natural selection is the basis of social instincts and thus of the development of the moral sense.

He thinks that humankind is poor and unemployed more or less as a result of an evolutionary process. "Light will be thrown on the origin of man and his history", he wrote in *The Origin of Species.* His later writings point in another direction altogether.

In almost all accounts since Darwin humankind is depicted with a split personality. On the one hand we are in concord with nature, being its product, on the other hand we are outside and a threat to that very same nature. Something seems to be lacking to make the picture coherent, reminding us of the way that Ptolemy had to insert epicycles and very advanced methods of calculation to make his construction of the solar system acceptable.

What is missing? Did Darwin and all his followers overlook something? Ptolemy's view of the world was inaccurate because he overlooked the most important thing of all: that the earth is circling the sun. Maybe Darwinists miscalculate when they simply contend that "the human body is a product of a long evolution; it cannot be otherwise with our nature". (RE)?

When believers in evolution are confronted with their own picture of a human being who is both inside and outside nature at the same time, they do exactly like Ptolemy. They do not look for another possible explanation, but go on pursuing the same line of thought. They mend their picture with different tricks, not realizing that all tricks are in vain.

Why regard human beings as products in the first place? It makes me think of cars leaving a factory or planks coming from a saw-mill. Is humankind comparable to that?

The word "product" evokes associations to something that we will eventually know everything about, a thing or an item produced by someone. Are we such "things"?

Most people, I presume, would say that we are not. Why then does the idea exist that we are to a hundred per cent products of nature? I think it is a prerequisite of science. In order to comply with the standards of science we have to feign that we are some kind of objects able to be examined in their entirety. If we are involved in our own evolution, partly producing ourselves, things get very complicated. We do the same thing with life, knowledge, language and the universe itself: reduce, limit and turn into objects. By limiting, by making ourselves into objects, we believe that we can grasp ourselves, achieve a tangible identity. The opposite might well happen: we turn a blind side to the true origin of ourselves.

To sum up: We cannot possibly be an integrated part of nature and at the same time its worst enemies. There must be something wrong with such a description. I believe the understanding of our role in the evolution of ourselves is missing. The idea of humankind as a mere product of evolution is dangerous. I will lead to passivity, to the conclusion that nature itself is to blame. "We are made in this way and can do nothing about it."

The second floor of the "house in cosmos" is as shaky as the first one.

Chapter 4
The magic of language

The next floor of the "cosmic house" deals with language. In order to understand our house we must know the function and origin of language. Most discussions of this aspect are Darwinian. So is Edberg's. The mere fact that language has evolved must imply evolutionary advantages, he argues: those who talked had more success in spreading their genes than those who did not. The ability to hunt efficiently was also enhanced by language.

"Language widened the living space of humankind" because we could memorize and plan, record history and pass on experiences. We used words to survey and sum up.

But language also came with a considerable drawback: it blunted direct and unobstructed experiences, cutting our navel string with nature. "The fragile and delicate part of an experience, what we want to preserve, is shattered by the blunt instrument of words."(RE).

Words are suddenly compared to the bars of a prison through which we view nature. Symbols are regarded as intermediaries to physical reality. In the same instant we started to use words we were expelled from Paradise. In Edberg's view a high price to pay.

This proposition is strange. We are first convinced by the author that experiences were made possible by language. This is followed by an assurance that there were experiences *before* language. This strategy is repeated throughout the language discussions in the books of the author: the prerequisite of having a language seems oddly enough to be – language.

The question of the evolutionary advantage of language has tormented biologists since Darwin. If no such advantage is found the whole theory is jeopardized: language could not have developed without evolutionary gains. Scientists of all sorts are in on the hunt.

The result is strikingly meagre. Some authors stress the improved ability to cooperate in groups while hunting, just as Edberg does. Others disagree. I think Susan Blackmore has a point when she criticizes the hunting-theory in her book *The Meme Machine.*

> What is not quite clear from any of these theories is why humans, and humans alone, should have developed such a complex and neurologically expensive solution to the problems of hunting or foraging. For example, wolves and lions achieve clever pack-hunting strategies without grammatical language, and bees communicate the whereabouts and value of food sources with a specialised dance.

Others Darwinists, such as Stephen Jay Gould and Noam Chomsky, think that language is a by-product, caused by what Darwin called "pre-adaptation", which is now known as exaptation. This is the idea that a species uses an adaptation for a purpose other than what it was initially intended for. The increase in brain size might for example have caused language to develop. Susan Blackmore disagrees: "It is no explanation at all to say that it came about as a by-product of something else or entirely because of physical constraints."

Robin Dunbar argues that language brought about social advantage, and might have developed from the grooming among monkeys. Some scientists seem to turn their backs on the whole problem in a manner similar to what happened in France in 1865, when the *Société de Linguistique de Paris* forbade its members to discuss the origin of language. They simply established, like Neil Carlson, that "the ability of verbal communication gives our species considerable advantages". Exactly what advantages we are never told.

Susan Blackmore eloquently describes the problem in *The Meme Machine*:

> If the human language capacity is a biological system like the vertebrate eye or bat sonar then we must be able to say what function it served, and why individuals carrying the genes that increased language competence were more likely to survive and reproduce than their less language-competent neighbors.

Darwinists of all kinds are eagerly trying to find the answers. Terrence Deacon, Robin Dunbar and Steven Pinker all think that either we must understand what selective advantage language gave early hominids, or we must abandon any hope of a Darwinian explanation.

I think their dilemma is caused by a limited view of language and consequently of humankind itself, a misunderstanding that leads them in the wrong direction.

According to Susan Blackmore we have to include a second replicator, the memes, to be able to find a Darwinian answer to why language exists. It remains to see if she is right.

How did language emerge in the first place? Edberg tries to answer the question, inspired by renowned predecessors. One of them is the Bible. Genesis seems to tell us all about it: humankind gave names to everything in heaven and on earth. Language thus arose through designation. Another predecessor is Augustine who confirms the Biblical idea in *Confessions.*

Edberg is just one in a long tradition using the same explanation. Language, he says, is name-giving, designation. The first sounds of language, he contends, developed when the hominids "started to imitate natural sounds and used them to designate".

These ideas are in no way new. In the few pages Darwin wrote about language in *The Descent of Man* there are many references to imitation. Darwin, of course, does not want to make a clear distinction between animal and human language in the Chomsky way. He agrees with an archbishop (!) who said that man "is not the only animal that can make use of language to express what is passing in his mind, and can understand, more or less, what is so expressed by another". Darwin also points to the inarticulate cries and facial expressions of human beings to link us to other animals, conceding that "the habitual use of articulate language is, however, peculiar to man". The difference is a difference in quality only.

"Man has an instinctive tendency to speak, as we see in the babble of our young children", he continues. Language "owes its origin to the imitation and modification of various natural sounds, the voices of other animals, and man's own instinctive cries . . ." Musical cries in sexual selection may also have contributed to the rise of words "expressive of various complex emotions". Maybe apes imitated the "growl of a beast of prey, and thus told his fellow-monkeys the nature of the expected danger?" Darwin concludes that "this would have been a first step in the formation of a language". He is convinced of his theories to the extent that he can say “we can trace the formation of many words further back than that of species, for we can perceive how they actually arose from the imitation of various sounds”.

Many philosophers and authors believe that the first human beings, before they started to talk, lived in speechless harmony with nature, in an “intangible world of impressions and sensations . . . a wordless world of conceptions.” (RE). Out of this human grunting and screaming the first words were articulated, with all probability as imitations.

The expression “wordless conceptions” seems to imply that the primitive being could see differences in his environment but not really tell himself what these differences were. How is it then possible to discern them? It seems as if the belief in “wordless conceptions” presupposes the existence of language, as if there was “a wordless world of words”.

Many authors regard language as “humankind’s own innovation”. (RE). But it can hardly have been invented like wrenches and violins. Wittgenstein saw it as a “graft on instinctive behaviour” and said in *On Certainty* that “language did not arise through reasoning”.

In other words: there is no thought behind the thought, no word behind the word. The virginity of language is absolute. There are no notions preceding it, as Edberg thinks.

If we look a little closer at the Darwinian idea that early man “started to imitate natural sounds and used them in designation”, we will find that this cannot be the starting point of language. It seems to me that the Bible, Darwin, Edberg and others who cling to the designation theory explain language “backwards”. They think that there is a wide coherent world to designate, but there is none. Think of the word “grasp”. Animals do this instinctively, without connecting it to anything else. It is a singular movement. What happens when “grasp” is connected to language? It completely changes and revolutionizes the world without changing anything at all! To grasp has ceased to be a singular event. It is now connected to other movements, to food, building-material, weapons, other people, to understanding. Suddenly a whole world appears out of nothing. The singular has become a coherent universe, a universe held together not by gravitation but by language.

It is impossible to designate the world when there is no world, just singular unconnected events. Language makes the world exist even though language changed nothing.

Therefore natural sounds can only be used as words if we already have a language. There is no point in doing it without one. If we lack language, how do we connect the sound of the stream to the stream itself? In other words: language is primary. There is nothing behind it. It stands on its own feet not as something added to the world, but as a part of the world.

The rise of language and human consciousness is far more complicated than the Bible, Augustine, Edberg, Darwin and many others imagine. Ludwig Wittgenstein, the Austrian philosopher, has given us all the keys. His *Philosophical Investigations* is largely dedicated to criticism of the name-giving theory. Language cannot have come about with humankind in full control of it, acting like a conqueror who sets out to name his surroundings. Neither can humankind have acted as an imitator who picks up different sounds from his environment. There never was this separation between humankind, language and world. The idea that we invented language is faulty. It is equally true that language invented us.

Designation cannot be a cornerstone of language, according to Wittgenstein. A name means nothing outside of its context. "The common opinion of the importance of words clouds language-function in a mist that makes clear-sightedness impossible."

His prime example is that of a complicated mechanism, where the "lever", the names of a language, is nothing without the other parts. "Separated from its support it is not even a lever; it may be anything, or nothing." We can compare these "solitary words" with the archeological discovery of a metal disc. What is it? How was it used? It is impossible to say without knowing its proper place or the culture of the producer. "This is the king" is an incomprehensible statement if you have never seen a chess game or heard of monarchy.

"Our knowledge forms an enormous system. And only within this system has a particular bit the value we give it", Wittgenstein says in *On Certainty*.

A similar idea is even more lethal to the theory of designation, and not only to that theory but to all theories of language. It is something that is never talked of, something that professors of linguistics pass by without noticing. I have not seen it mentioned anywhere.

Noam Chomsky is not aware of it when he discusses language-studies.

> No one would take seriously the proposal that the human organism learns through experience to have arms rather than wings . . . Rather it is taken for granted that the physical structure of the organism is genetically determined . . .
>
> The development of personality, behavior patterns, and cognitive structures in higher organisms has often been approached in a very different way. It is generally assumed that in these domains, social environment is the dominant factor. The structures of mind that develop over time are taken to be arbitrary and accidental; there is no "human nature" apart from what develops as a specific historical product . . .
>
> But human cognitive systems when seriously investigated, prove to be no less marvelous and intricate than the physical structures that develop in the life of the organism. Why then should we not study the acquisition of a cognitive structure such as language more or less as we study some complex bodily organ?

Yes, why? Why is it impossible to realize the hopes of Chomsky? We will see why if we ask ourselves what is used by all scientists in all studies of language. Not a single study has seen the light of day without it. What are the professors, the learned men, doing? They think, talk, lecture, discuss, make speeches, write books and articles. *They use language.*

In other words: they are using the "organ" they examine when they explain it. And they cannot in any way leave the situation by pointing at the language on the pages of books or in recordings. What they use to understand the object is still the same object. They are high-jacked by language and not at all the free agents they imagine themselves to be.

They are parts of a system when they try to explain it. And that spells having difficulties with the explanation. If you do not agree with me you should believe the mathematician Kurt Gödel who showed that an axiomatic system cannot be proven as reliable, not be understood *from within itself.* To examine language with language is tantamount to measuring a tape with the same tape or examining the liver with the same liver. You get a result that might seem convincing, but it is only provisional and will be succeeded by other temporary results. The exact length of a measuring-tape cannot be established by measuring the tape with itself.

What does this mean? It means that we will never be able to answer the question "what is language?" It is simply not an object in the way we think it is, an object open to study like other objects. Looking at it in the Chomsky way we dismantle it and partly "kill" it.

What is language? The only answer is, as I see it, offered by Wittgenstein.

Language developed in a symbiosis with nature and human consciousness, not a regular symbiosis like those we find in nature, and not an easily detectable one, but still a symbiosis.

Wittgenstein is not the only one who points in that direction. So does Jean Paul Sartre when he says: "Consciousness and the world are given together." And language!

What does this mean? It means that world and language communicate, that the essence of the world is the essence of language. It also means that language has separated us from our surroundings without doing so, that it designates without doing so, that it looks like a separate "organ" as Chomsky thinks, without being one. *Language is something that in itself preserves the symbiosis, even though this preservation is not obvious or easily detected.* This is why we can say that language dropped an atom bomb on paradise without stirring a leaf.

The animal who became human lived in harmony with nature. I think we can agree on that. Did human consciousness and language really put an end to that? Or is language just a continuation of an unbreakable harmony, one that we have tragically overlooked?

Genesis shows us a paradise where the first human beings are lured into eating the forbidden fruit. If this fruit is language and human consciousness, it must mean, if the symbiosis is unbroken, that Adam and Eve were exiled from paradise without being so. Language changed everything without changing anything at all. But still the banishment is real. It is an experience lived by millions of people. Who does not feel the pang of sorrow when reading about the exile in the Bible or listening to the lines of John Milton?

> They looking back, all the eastern side beheld
> Of Paradise, so late their happy seat,
> Waved over by that flaring brand, the gate
> With dreadful faces thronged and fiery arms:
> Some natural tears they dropped, but wiped them soon;
> The world was all before them, where to choose

> Their place of rest, and Providence their guide;
> They, hand in hand, with wandering steps and slow,
> Through Eden took their solitary way.

This is an even more tragic occurrence if it never happened, or if it happened because we did not understand ourselves or our language – if *we are the symbiosis with nature with every word we utter.* If so language cannot only be regarded as "a creation by us". It is equally true that we are a creation by language. Wittgenstein is acutely aware of this. He often points out that language appears more like animal instincts than a tool or an instrument that suddenly was put into our hands. It might be as futile to ask "what is language?" as to ask "what is an animal instinct?" Can we imagine birds, wolves or cats without their instincts?

According to Wittgenstein language must be seen as "a graft on instinctive behavior". Even "the developed, complex use of language by adults embodies in its entirety something that resembles instinct", he writes. In *On Certainty* there are numerous references of this kind. "But that means I want to conceive it as something that lies beyond being justified or unjustified; as it were, as something animal." In other passages he is more detailed:

> But more correctly: The fact that I use the word "hand" and all the other words in my sentence without a second thought, indeed that I should stand before the abyss if I wanted so much as to try doubting their meanings – shews that absence of doubt belongs to the essence of the language-game, that the question "How do I know . . ." drags out the language-game, or else does away with it.

"Absence of doubt belongs to the essence of the language-game". This means that language is more important to us than we usually admit. It is the foundation of ourselves.

> I want to regard man here as an animal; as a primitive being to which one grants instinct but not ratiocination. As a creature in a primitive state. Any logic good enough for a primitive means of communication needs no apology from us. Language did not emerge from some kind of ratiocination. (LW in *On Certainty*)

If we cannot stand beside language and treat it as an object, or any "organ", if we *are* language more than we *have* it, how does it work "being language"?

I would like to call what language does to us as if behind our backs *predetermination.* It can be seen as something we stand on in order to stand at all. Or a first step that makes all other steps possible. The decision is already there when we decide, which makes it possible for us to make decisions at all. If we imagine a carpenter who is used by his hammer while he simultaneously uses it we get a pretty good picture of what language does.

Wittgenstein gives a beautiful account of this readiness of language.

> One day man opens his seeing eye, and there is light.
>
> What this language primarily describes is a picture. What is to be done with the picture, how it is to be used, is still obscure. Quite clearly, however, it must be explored if we want to understand the sense of what we are saying. But the picture seems to spare us this work: it already points to a particular use. This is how it takes us in.

"It already points to a particular use." This is what I mean by *predetermination.*

Wittgenstein has convincingly shown how predetermination works, especially in his discussion with G E Moore, professor of philosophy, on knowledge and certainty. Moore had tried to explain why the outer world is stable and what we can know with certainty in *Proof of an External World.* During a lecture Moore listed some of this unquestionable knowledge.

> I am right now, as you all can see, in a room and not outside, I stand up and do not sit or lie down, I am dressed and not completely naked, I speak in a relatively loud voice and neither sing, whisper or am quiet, I hold in my hand some sheets of paper with something written on them, there is a great number of other persons in the room together with me and there is a window in the wall over there and a door.

To this list of certainties Wittgenstein objected by wondering when Moore knew that he was dressed and not completely naked. Wittgenstein argues that Moore is mixed up in his view of knowledge, and that he has not clarified to himself what he means by it.

Knowledge, Wittgenstein says, is something we can doubt. It can be changed. New discoveries can be made that modify what was previously known. If anyone doubts that he is dressed or not, he does not doubt his knowledge. Such a person is out of his wits.

> To say of man, in Moore's sense, that he *knows* something; that what he says is therefore unconditionally the truth, seems wrong to me. – It is the truth only inasmuch as it is an unmoving foundation of his language-games. (*On Certainty*).

What Moore is talking about is something that is a foundation of knowledge rather than knowledge itself, something beyond doubt, language as "a graft on instinctive behavior".

Wittgenstein points to the *predetermination* of language, even though he does not use the word. And this is something unknown to Moore, a part of language he has not yet discovered.

Why then all the certainties on Moore's list? Because they are there for other reasons than to explain knowledge, as a pretext. Remember the words of Wittgenstein: "This is how the picture takes us in." Everything is already said when Moore tries to say it. It is prepared. Done. Then Moore reflects on it afterwards and takes the reflection as proof of knowledge. That is why Wittgenstein characterizes Moore's "proof" as a kind of tautology. I would add that it is an attempt to wrestle from language what ultimately belongs to it and not to us. Moore's claimed certainties can to a certain extent be compared to stolen property.

That is also the reason why Wittgenstein so often criticizes philosophical thinking in general. "I am sitting with a philosopher in the garden; he says again and again 'I know that is a tree', pointing to a tree that is near to us. Someone else arrives and hears this, and I tell him: 'This fellow isn't insane. We are only doing philosophy." (*On Certainty*).

What is the essence of Moore's demonstration to his students? It is denial. He denies the power language has over him in the form of predetermination and claims all power for himself. The grabbing of power is the hidden purpose of his denial. But what seems to establish the human kingdom is exactly what hampers and troubles it and ultimately may lead to its destruction. It is the act that brushes aside the symbiosis between ourselves and nature/language. Alienation is born in this instant, created by ourselves as behind our own backs. To our great misfortune this alienation has been called existential as if we had nothing to do with it. Moore, and all with the same idea, open the box of Pandora. In one blow we lose what we cannot lose, fail to secure the concord with nature and become alienated.

What people call "human nature" now flies out of the box like an untamed horse; sowing hatred, jealousy, pride, greed, restlessness, anxiety, fear, depression and despair.

The certainty that Moore aimed to demonstrate in the first place is instead the cause of uncertainty, and a double one at that: one that has to do with our identity and one that has to do with the identity of the world. The real, hidden certainty is to be found somewhere else, in the discovery of the necessary predetermination of language, in our being conscious of the fact that we are talked when we talk, that we are indestructibly tied to language.

What Moore calls knowledge should be regarded as a predetermined word or a word that he is. It is exactly what Wittgenstein points out: something he cannot doubt. In the same instant as Moore neglects to recognize himself as being language rather than having it, he severs himself from the only connection with nature there is, i.e. instinctive language. This makes it impossible to take full responsibility for himself or his surroundings. Instead he will believe himself to be secondary, regard himself as created, be it by God or the evolution. What he will gain from this is an identity with the inner core of a mystery.

Now we are close to answering why environmental work is so tardy. *It is because we have not localized our true connection with nature.* If we are language more than have it, we are primarily linked to language in the form of nature rather than to nature itself. It means that our search for "an ecological link to nature" is an empty gesture. There is no need for a search because the connection is already established by language. We cannot even get rid of it.

It is deeply tragic that Moore's standpoint and not Wittgenstein's is the dominating one in everyday conversations, in ordinary philosophy, in schools and at universities.

So when we want to find out about "the relationship between language and reality" this endeavour is also part of the misunderstanding. On *YouTube* the philosopher John Searle uses those very words in his first lesson on language. This expression is used by many linguists. But language as I have shown with the help of Wittgenstein is also reality. If you talk about "the relationship between language and reality" it is the same as talking about "the relationship between reality and reality". It simply does not make sense.

But what about our imagination? Does it not liberate us from predetermination, making us carpenters that are not nailed by our hammers when we use them? The evolutionary psychologist Robin Dunbar has suggested that what makes us different from monkeys and apes is the ability to imagine. We can live in virtual worlds, he says, due to our enhanced intellectual capacities. We can make plans for the future and "imagine things to be different from what they are." Other species cannot do this because they lack the ability to leave reality and ask themselves how everything would look if it was different. Is this really a valid description of our capacity to imagine? I think not. We depend on the predeterminations of language which we cannot be without. They make imagination possible in the first place.

We can only imagine a different life or future within the framework of our predeterminations. Or can Utopians describe what life in a society of equals would be like? What kind of language people speak there? What kind of world view is theirs?

Utopians are not to be trusted because they omit the existence of predeterminations. They think they are at liberty to fantasize about the future in the way Robin Dunbar describes. This is not possible. All the imaginations of visionaries and Utopians are determined by a language that in turn is determined by the social order where the language develops. What Utopians give us is just a rearrangement of what is already known. They juggle reality – do not renew it. It might look new to the naive eye but the predeterminations are there beneath the surface of the Utopian fantasy. In order to think what no one has thought we have to change our predeterminations. This is very difficult without living in a different social order that makes such changes possible. But the process can start if predeterminations are brought to the surface and scrutinized, if you get familiar with the thought of being language.

The predeterminations of language are hidden in what is regarded as self-evident. The scientific idea that the world needs enlightenment is a good example. It seems self-evident and unquestionable. We do not know everything and would like to know more.

Enlightenment started in the 17th century with philosophers such as Francis Bacon and Rene Descartes. The idea was to leave the superstition of the earlier centuries behind and correctly read the pages of nature in the book that God himself had put in our laps. The darkness was to be dispersed. But who created the darkness in the first place?

The answer to this question is not easy to understand: I think we did unknowingly, through language predetermination. Language and consciousness fixes our eyes in a position, as with the strapped down persons in Plato's cave. We are, in a sense, prey to language.

The whole process looks something like this: We are positioned in a room, where we close the window shades without knowing it. These are the predeterminations. Then we leave the room to enter it again. Now we think that we are entering the room for the first time. We discern things as vague and obscure and try to enlighten the dark room.

The scientific approach seems axiomatic. We seem to be placed in a world that to a large extent has the characteristics of a riddle, a mystery. We can put this in another way: By first making the world into an object we mystify it, move it out of range of real understanding. If the world, language and human consciousness are given together we cannot separate one from the other without destroying the interrelation, the symbiosis. Christians misinterpret this separation calling it "original sin". But there was of course no sin involved.

Many authors and philosophers have tried to describe this need of Enlightenment. A well-known Swedish author as well as an American citizen, Lars Gustafsson, writes that "we are parachutists landing in a foreign landscape, in complete darkness". He calls this notion "an existential axiom that all philosophy students have to learn". Another Swedish author, paints a similar picture in a book titled *The Riddle of Consciousness.*

> I come to think of the ice floe that broke loose with a group of Estonian-Swedish seal-hunters and was adrift on the Baltic Sea in the winter of 1796. The hunters could walk about on it, hit it with their heels and feel the firm resistance, but beneath and around the floe there was just impenetrable dark water. (J L).

These pictures show us as strangers totally unconnected with the environment. If this is our situation, if we are parachutists landing in a foreign landscape or people trapped on a slab of ice, what does that do to our understanding and our environmental responsibility?

Why should a stranger care about what he did not create in the first place, something that he is not even part of? If darkness is the initial word, what then do our words mean?

A parachutist is a person in passing. He comes from a place other than where he lands. With such alienation it is virtually impossible to find a stable human identity.

Daniel Dennett, the American philosopher, is therefore another advocate of permanent alienation. In *Darwin's Dangerous Idea* he thinks that "the real world is always hidden from us like a needle in a haystack". Is it? And if so, will not every scientific advance remind us of this "fact"? What happens to the hope of Enlightenment if the world is always hidden?

Stephen Hawking follows the exact same path when he says that "we live in a puzzling world. We want to find an immanent order and ask ourselves: What is the nature of the universe?" But that question, like others of the same kind, will never be answered. It just makes alienation more persistent and more severe. Hawking's idea is but another predetermined notion, growing out of the lack of awareness of the language process.

If we are determined "in advance" by language how can we influence that process? Is it possible to change the predeterminations to make them work with us instead of against?

Yes it is. And I will show how in the next chapter.

Chapter 5
Language and society

Where do predeterminations come from? Why do we think that we are secondary, that we have been created by God or that we are products of the evolution? Or why do we embrace other related, prevalent notions, such as belief in destiny, myths and our own frailty?

I think the answer to those questions is quite obvious: because such notions make our type of society sustainable. They are the oil of its machinery, tailored to fit the way we live.

What is the predominate structure of our society, one that is always present, emphasized in schools and institutions, in family relations, male and female roles? It is by far inequality, established by means of dominance and submission: the power distribution.

During a very long historic time our societies have rested on that pillar. Since infancy we have been accustomed to roles of dominance or submission, and obtained much of our identity from these roles. This identity is mirrored in our bodies, our language, our gestures and posture. It is more than likely that the predeterminations grow from this order.

Inequality is an evident and very tangible structure of our societies. Still, it is obscured or even denied. We try to conceal it with all kinds of myths and ideas telling each other that "we are created equal", "we share the same human nature" etc. But the fact is that people are separated at an early stage through different mechanisms, schools being one of the most important. We are also separated by gender, as men are favoured in the labour market with higher salaries and women often occupy lower positions. The labour market also puts a price-tag on everyone who enters it, which obscures our real value. It is not surprising that in such a society we enlarge and belittle ourselves, that self-esteem is a frail commodity.

Getting socialized to inequality and to a language that preserves hierarchies can be described as a "work of transformation that is the more powerful as it is to a great extent invisible and stealthy, through imperceptible habituation to a symbolically structured world and early, durable experience of interactions permeated by relations of dominance".

"Invisible and stealthy" says the French sociologist Pierre Bourdieu. When he examined how this largely invisible transformation works, he compared it with music and sport.

A skilled musician can easily improvise within a given harmonic structure without having to mentally repeat the variations in advance. In the same way as a musical "language" is "innate" with a musician, inequality with its predeterminations is "innate" with us. But there is a significant difference. Inequality is all around us. That is not the case with music. A musician can get up from his chair and leave his instrument to live in a silent world, but we can never leave the order we have inherited. It is as present as the air we breathe.

If you live in a society of hierarchies where you either dominate or are dominated or play both roles, this fact ought to be reflected in your view of yourself and the world.

What we have just discussed, the belief that designation is a cornerstone of language, is in my opinion such a reflection. The idea could only arise in an unequal, hierarchical society. The theory of name-giving regards the name-giver as someone who is separated from his environment, conquering it, showing his power and his ability to master the world. He or she is at home with this dominance only because they breathe the air of inequality.

The manner in which we view the world originates in the same way. Religion "imitates" the hierarchy of society by regarding us as secondary creatures. We are subordinate to God who is our superior. Being equal to God is an absurd and blasphemous idea. Instead God is regarded as elevated, a harsh judge and also a mild and forgiving father. The story of God in the Garden of Eden mimics the patriarch's role in the ancient society where the text was written. God makes all the important decisions. He gets angry when not obeyed. He is infinite and perfect where the human beings are finite and imperfect. All these ideas stem from the inequality that runs in our blood, from the "imitation" of social structures.

Those who are religious do not unmask this veil of ideas but think that it is self-evident that God is our superior. They do not see the link between language and society.

Not even science, which is supposed to disperse the darkness of the parachutists, can manage without superiority and inferiority. It tells us that we are subordinate to nature, that Mother Nature created us and gave us our identity. Nature is primary and we are secondary. This has become an axiom of the modern era. Science, regarded in this way, is the rational "imitation" of religion. God is substituted by Evolution as nature's creative principle.

In my opinion the notion that human beings are subordinate to God or Nature fits an unequal society as water fits the bowl. Ideas that belittle human beings are easily adopted in such a structure. Language and world view prepare us to accept powerful leaders, even tyrants and dictators. We look to leaders and are prone to give up ourselves for abstract constructions such as money, progress, political ideas or technological advances. Again we do not see the link between society and language. This is because the origin of the ideas that uphold society is hidden by the same process that produces them. Or as Bourdieu puts it: "The system as such produces the concealment of the truth about the system and its effects."

We do not know how people in a society of equals would view themselves or their environment. We do not know what their language would be like, what meanings their words would carry. But it seems unlikely that they would regard themselves as inferior or superior to any other human being. They would surely have another identity than we do, look upon themselves differently. Would their language be different? Would we understand them?

They might well be the lions that Wittgenstein referred to in *Philosophical Investigations*: "If a lion could talk, we could not understand him."

In a society language is infected by predetermined ideas that fit this society. Those who try to detach themselves from these ideas run into difficulties. If they manage to find parts of a new language, it might even be unintelligible to others. Language and society raise a joint barrier against those who seek to alter or exchange the predetermined notions.

Wittgenstein was aware of the difficulty and describes it in *The Big Typescript.*

> People are deeply imbedded in philosophical, that is, grammatical obscurity. To liberate them means to detach them from the immensely complicated bonds that captivate them. You have so to speak to regroup their whole language. – But this language has become what it is, because people have been – and are – prone to think like that. Detachment therefore is only

> successful for those who are instinctively revolting against language. Not for those who with all their instinct belong to the herd that created this language as their proper expression.

Here Wittgenstein is very outspoken about the predetermination of language, how we are attached, and how this attachment builds identity. What is needed in order for us to be liberated, and also for us to understand ourselves and rid ourselves of the socially created notions, is a revolt against language. It goes without saying that he means a radical change, one that does not take place sporadically or at certain times, but one that permeates every part of our lives. The whole language needs "regrouping". Still Wittgenstein, in my opinion, partly misses what he aims for. There is some elitism and idealizing here that complicates the solution of the problem. But we have to consider that *The Big Typescript* was not intended for publication. It is more like a philosopher's workshop, unfinished bits and pieces.

It is very easy to misunderstand what Wittgenstein says here, that is, if we are unaware of the predeterminations and imagine we are free agents vis-à-vis our language.

This is also exactly what Georg Henrik von Wright did, the Finnish philosopher who succeeded Wittgenstein as professor at Cambridge. von Wright administered Wittgenstein's papers after his death and had no objections to publish *The Big Typescript.*

He quotes the passage in his book *The Myth of Progress* and misunderstands the whole thing by thinking that Wittgenstein is talking about a simple sanitary procedure.

> There are in the vast garden of language well-kept patches where untainted language games are played by human beings. But this garden is also partly overgrown by metaphysical weeds that obscure a clear view of the patches, erase their borders and by doing so confuse those who play the games. The good philosopher should be a gardener who by weeding uncover language areas in their purity and thus helps communication to continue without the disturbance of metaphysical-philosophical confusion.

This is not at all what Wittgenstein meant. He talks about a vehement instinct, von Wright about a rational gardener. Why the instinct? Because it is almost impossible to grasp the "new" language inside the old one, to be part of a society that does not yet exist.

von Wright thinks that everything that is needed is already there. All you have to do is to pull up a few plants and water others to be able to play what he calls "untainted language-games". Society that gives language life and value is nowhere to be seen. It is not involved. Neither does it involve people in the core of their being, which is exactly what Wittgenstein's theory does. According to von Wright people only have to be a little more meticulous in order to play the "pure" language games. The only thing that has to change is language and as we have seen – and as I have tried to clarify – language predetermination cannot change unless society does. Gardening in the hands of von Wright is a simple procedure. To Wittgenstein it is something much more difficult: it is rather a question of removing all the plants, the soil and the solid rock underneath: the totality that creates our predeterminations.

Who plays these "untainted language games" that von Wright envisages? No one. von Wright opens the door wide to elitism, divides humankind instead of unifying it. But so does Wittgenstein. He partly has himself to blame for von Wright's misunderstanding. I can see why he was against publishing *The Big Typescript.* His statement is not a finished one. The elitism is obvious, when people are identified as a "herd" and divided into different categories: the wise who instinctively revolt and the sluggish who do not.

If we erase elitism from his statement it is acceptable. "But this language has become what it is, because people have been – and are – prone to think like that." Why do we think like that, willingly carry the burden of predeterminations on our backs? Because of the environment, the structure of society. I think Wittgenstein is fully aware of this.

There is another point to be made here. If von Wright is correct, (and partly Wittgenstein), those who "created this language" must be blamed. And in blaming we forget that they are also victims, albeit of their own doing. If we omit this aspect, we also omit compassion and empathy. Can you feel compassion for a neglectful gardener? Who weeps over carelessness?

If we go on analysing this passage of *The Myth of Progress,* we see that von Wright loses his way even more. He talks about the possibility of an "uncorrupted language" that will give "a clear view of the true character of things". He treats the possibility like a vision, doubtful of its realisation.

The misunderstanding looms above his head, larger than ever and is embedded in the belief that there can exist "a clear view of the true character of things".

von Wright does not understand that the "things" never will show their "true character" through science (or religion). This true character is not attainable by an examination of reality as an object which Wittgenstein was well aware of and several times expressed as the sum of his philosophy: "What belongs to the essence of the world language cannot express."

Unfortunately most philosophers do not take these words into consideration. It is certainly a deficiency with von Wright and countless others. Whoever tries to express *the essence* of the world, or any *true* property of the world, does not express *it*, but something else.

When we do what Wittgenstein recommends, examine how language is used in different situations and contexts, we soon discover its dependency on domination and subordination.

I have already mentioned the relative easiness with which powerful leaders and even dictators are accepted. In communism the Party was superior to the individuals. But a similar thing, making people subordinate, goes on when Economy or Development are regarded as some kind of superior powers. It is easy to forget that we are the only prime subjects. Our whole language seems to point in the other direction, making us secondary. Earlier in this book I discussed the expression "from the viewpoint of the stars". There are numerous expressions like that, for example "from a biological perspective". They all conceal our role as the sole responsible subject. Richard Dawkins does the trick when he puts the "selfish gene" in the subject's place. And brain-experts do the same when they talk about the brain as a self-governing subject, looking at the brain as if it was detached from our bodies.

To a great extent domination and subordination decide the value of human beings, which make some people think that they are more worth than others (or less worth). As I mentioned self-confidence is easily inflated or deflated in such a society. We are constantly told by philosophy, art, literature, film, religion, science and all kinds of media that we are weak, imperfect, faulty, guilty, sinful temporary creations that are here by chance, and that it is human to err and waver. We have in fact been compared to sheep in our tendency to be seduced by all kinds of siren calls. We are told that we cannot know everything, that life is a mystery and an ordeal and that we all need comfort, forgiveness and indulgence. But all these notions do not originate with "human nature" but are products of the way society and language work by reinforcing each other, creating a breeding-ground for such notions.

We should not be surprised that individuals who regard themselves as weak, subordinate, secondary, thrown into a vast mystery where they sooner or later will no longer exist, have a need of a superior power, a deity to lean on. But this need comes from the removal of the true subject-role. If we were all primary subjects the need for religion would disappear.

Our not being complete subjects creates a vacuum that has to be filled. The human identity has to be supplemented by something. Consumption, power, ambition, status, greed and many other dispositions are rooted in a sense of need and imperfection. This is how people are exploited. The road is paved for those who create artificial needs.

The lack of a stable identity explains why so many people represent something other than themselves, for example God, humanity, the state, art, ideologies and truths of different kinds. It also explains why there are fanatics and fundamentalists and why many people aspire to become celebrities. It is as if we are not sufficient in ourselves but have to look for our fulfilment somewhere else, in some label, faith, enterprise or undertaking. And there always remains the everyday trick to talk about "humanity", "we" and "us" to fill the gap.

Dominance and subordination are endemic to our societies. It is part of the social structure and as such permeates our language. But the influence does not stop here. Our bodies are also affected. "The subordinates", Bourdieu writes, "pay a tribute, often unaware and against their own will, to their own subordination by silently accepting the prescribed borders". This tribute, Bourdieu continues, "often takes the form of *bodily emotions* – shame, humiliation, shyness, anxiety, guilt – or *passions* and *feelings* – love, admiration, respect; feelings that get more painful by visible effects like blushing, difficulties to express oneself, clumsiness, trembling, wrath or impotent rage". I believe that all who have been dominated or who dominate will find at least some trace of this in their speech, bodily posture, gestures or facial expression. Just compare high-positioned persons with those in the lower ranks!

Language has its roots in our society and our circumstances. What we say is really part of what we do and a key to the identification of ourselves.

Our type of language makes us think that it is self-evident that we have been created, either by God or by the evolution. But language also tells us something else, something that is hidden from a direct view. Through its power of predetermination it whispers to us that it has no beginning or end, ruling out any attempt on our part to find its origin or view it as an object detached from ourselves. If we try to examine language in that way, we ruin our chances of a final understanding.

We not only have language – we also *are* language. And understanding this in its profundity is in my opinion the same as understanding our concord with nature. The identity we have looked for with the help of religion, science, ecology, art and literature, is already there through language, established with an unsurpassed power and brilliance.

Chapter 6
A final survey

Among scientists there is a widespread belief in the unifying "theory of everything" even though the enthusiasm of the past decades seems to have dwindled. A final theory, a profound combined knowledge, is deemed necessary for us to rightly manage our "house in cosmos".

> What we need is a comprehensive philosophy that gathers all knowledge – from astronomy to evolutionism, ecology and ethology, genetics and psychology – making an overall picture of the universe, life and our own destiny. (RE).

Almost the same words are used by today's ecologists and environmentalists.

> Making people aware of our concord with planet earth will demand efforts from science, culture, philosophy, and religion. These efforts must be monitored by a moral compass of responsibility between generations and esteem of nature. (JR, MK).

None of these authors, and very few of those I have read, seem to question the reliability of the combined efforts of science, culture, philosophy, and religion. And yet, all this knowledge is underpinned by the predetermined notions and ideas of an unequal society in the way I have just described. Very few authors ask themselves if this overall view will help them or lead them astray, or if it is at all possible. They trust in its existence.

Philosophy, of old, is divided on the issue. Some philosophers, like Hegel, think that an overall view (or system) is possible. Others, like Kierkegaard, rule out the idea. Maybe the whole history of philosophy can be looked upon as a duel between these two ideas.

There might well be a system, Kierkegaard says, but none of our existence. And Jean Paul Sartre, his existentialist disciple, agrees: "A totality is in principle unattainable."

It is the hope of many writers of biology and ecology that science and culture will find this totality of knowledge making it a link to nature. It was the cherished goal of those who started the fight against pollution, pesticides and other dangers in the 1950s and 60s. What has happened since then? If we take into consideration the flood of information on the issue, the knowledge and awareness, we would have taken giant leaps. Scientists say that they know exactly what measures we should take. And if you read a newspaper, listen to radio or watch TV, you also get a lot of information. Has this information and knowledge changed the world in proportion to its abundance? If the answer is negative, why has it not?

Could it be because we are robbed of our position as sole subjects, apt to look at ourselves as secondary, as created beings and thus not fully responsible for our actions?

How can "a product of the evolution" (or someone created by God) be fully responsible for itself or other living creatures? This product will have a nature given to it by the evolution (or God). Ecologists like Johan Rockström never tire of telling us that "we are in control", "we are in the driver's seat" and consequently able to mitigate the wounds of nature. But if God or the evolution instilled a human nature, the total control is out of our hands.

Then *we do not* sit in the driver's seat. God or Nature does. Edberg and Rockström are not alone in accepting this view. Scores of authors, philosophers and scientists do it when they proclaim that we are products of the evolution, among them Richard Dawkins, who robs us of our subjectivity by proclaiming that we are mere vehicles driven be a "selfish gene".

Charles Darwin is of course behind it all by mixing us up with animals. Evolution gave animals instincts and a nature. Why regard humankind differently?

A unifying theory will probably never see the light of day. That does not mean that we must stumble when it comes to caring for our "house in cosmos". On the contrary, I would say. The search for a total Enlightenment, might even be counterproductive ecology.

Scientists have a tendency to mortgage the future. And they can do it undisturbed because no one is able to call the cards. They say: it is not here today, but we will probably have the full picture in a few generations. Mortgaging the future is a deeply rooted habit.

If you study environmentalists they are all smitten with this malady to go beyond present knowledge and predict the future. They also have a tendency to push the boundary of "the darkness around us" by making their knowledge more reliable than it is. They do so when they on the one hand put themselves in a godlike position in expressions like "from the viewpoint of the stars", "from a cosmic point of view" and on the other hand shrink to almost nothing when they admit that they are not omniscient. "Our knowledge can be but fragmentary." (RE). "If we knew everything would we not lose the excitement of life?" (RE).

The habit of scientifically investing in the future, predicting results that might not be there, was started by the philosopher Francis Bacon, also called the Father of Science. Nature is but God's open book, he said, finite and thus legible from the first page to the last. In due time we will understand everything because God will never betray us.

Three decades ago, Stephen Hawking declared that a "theory of everything" was on the horizon, with a 50 per cent chance of its completion by 2000. Now it is 2015 and Hawking has given up. But it is not his fault, he says: there may not be a final theory to discover after all. Even so we should not resign. The riddles of existence might be explained without it.

I do not think they will be – such hopes also mortgage the future. Scientific development seems to go the opposite way: the more scientists know about the world the more complicated it seems. The simple models our science teacher showed us in the 1950s no longer suffice. Now scientists talk about dark energy, dark matter, antimatter, and virtual particles that exist too short a time to be studied. Cosmology seems to turn in the same direction with the idea of multiple universes. "The house in cosmos" is obscured by a cloud of theories.

I think there is more to every theory, and every use of language, than just theory. As I pointed out in an earlier chapter language has its roots in society, in our inequality. I think that the present language predeterminations turn our heads in a certain direction, making us look for things in places where they never will be found. This is the case with the origin of language. It is also the case with the origin of life, with our identity and with a unified theory. What we look for cannot exist with the predeterminations of today, because they make us look in the wrong direction, believing there are answers to questions that will never be answered. Even so, we should not resign, just try to turn our heads in another direction. It might well be as Wittgenstein pointed out, talking about humankind's search of truth: We seem to be in a locked room and climb to the window. But it is too high. We try the chimney, but it is too narrow. All the time the door has been open behind our backs.

I firmly believe in a totality of knowledge, but not the kind of knowledge presented by science, a perspective that objectifies the world and everything in it. It goes without saying that someone who is part of "the house" cannot separate himself from it to look at it as a mere object. This fact is not acknowledged by biologists, ecologists and other scientists. They simply brush it aside, if they are at all aware of it. They act as their colleagues the linguists who examine language with language, firmly believing in their strategy.

It is only through language, as I see it, that we can find our true identity. But that road is hard to find and painful to walk, at least to start with. It involves an instinctive revolt against our present language, like the one Wittgenstein sketched, a kneading of the dough of words to remove the airy emptiness. In *Tractatus* there is a guiding star: "Whereof one cannot speak, thereof one must be silent." We are on the right track, I believe, if we see the link between theory (language or mind) and practice (body, society). Language starts to blow bubbles when this link is broken. The ideal must be that theory and practice are Siamese twins.

The shift from an "empty" language to one that is involved in our own creation is not easily undertaken. *It is like walking in the rain with a bared head without getting wet.* We are immersed in notions and ideas that all serve to uphold the society we live in. The difficulty lies in dismantling these ideas. It seems as if they are the only ones that exist.

"Humankind is created by nature" ís such an idea. It seems to be a cornerstone of our identity impossible to replace. But that idea has, like all ideas, been given to us by a language dependable on our society. There must be such ideas and notions if we are to live at all. But that does not mean that *this* idea is beyond criticism. There can be others even though the difficulty to replace the ones we have is enormous. We are locked inside our language like prisoners in a prison and cannot get out, if we do not recognize our situation.

There are no exceptions to the rule that language is social. The words are not only words but also part of ourselves, the social flesh that we are. The question is not how we can find a totality of knowledge or a unified theory but how to replace a language which is a part of ourselves with another. How can we stand on a new platform when it does not exist? How can we change direction when there is no new direction? And still there has to be one.

To change direction we have to know the restraint. But that knowledge is scant. People think that they can say or do what they want, that there is something called "free will" and "freedom of expression". There is of course, but not in the predetermined sense I speak of. Karl Popper, the philosopher behind the idea of "the open society", obviously believed in those freedoms. But, as I have tried to explain, there are no open societies. All are guarded by unforgiving sentries, the predeterminations of language rooted in our social structure.

Popper thought that rationality, logic, and science all point to a society that is open and pluralistic. But in thinking so he misses the main point, how language has "prepared" the ground for us. He dreams of a society where opposing ideas are freely expressed, where everyone has a chance to examine problems and suggest solutions. But there is no such freedom, because there is a preordained ground, created by language/society.

Popper ought to have studied Bourdieu to discover that real change in a society is hard to come by: the quest for change very often preserves what it wants to change. This is also my point regarding environmentalists, but the same thing can be applied to politics or philosophy. The children of an unequal society tend to copy that society unawares. When Popper describes the authoritarian society he might as well be talking about the open one.

Those who govern the closed society lack the openness of democracy and the criticism of its citizens, he says. Therefore they "condemn themselves to pressing on with mistakes for some time after these have begun to produce injurious unintended consequences. The whole approach, characteristic of highly authoritarian structures, is anti-rational". Exactly the same can be said about our democracies and will be said in the face of the climate change. In my opinion Popper's distinction is not correct. It should be made, not between open and closed societies, but between open/closed societies such as the western democracies and closed/closed ones, such as North Korea – between "prisons" of different designs.

Popper has many kindred spirits in art and literature, people who think that they move around freely. The idea of modernism rests on this principle. It is a deadly sin to modernists to remain standing in the same spot. They have to constantly “renew” themselves.

This freedom is merely an illusion, because there are language predeterminations that direct our movements. Bourdieu is on the right track when he criticizes the liberal definition: “I doubt that there is real freedom except for the one that necessity makes possible.”

Marcel Proust, to mention just one of many, discusses literature in a way that reminds us of Popper. Reading to Proust is an intellectual sanctuary where people gain access to realities beyond their own reach. The reader can get his life changed without ever leaving his armchair. Is this true? If you line up *all* the authors you have a chance to read during a lifetime, they are born in unequal societies and have no other experience. You cannot read a book written in a society of equals, because such a society does not exist.

Can a writer imagine such a society? I do not think so. If that was the case we would have books in our libraries or in our homes with tales of such places and its people. To my knowledge there are none. Some people think that Utopians do the trick. I do not. All utopian ideas that I have come across mimic what they try to overcome. If you want to describe another society than ours you have to use another language. Who will understand it?

There are great differences between the authors you read during your lifetime. But this difference is at the same time a uniformity. In all prisons there is a space where you can move around freely. In a sense diversity can flourish. But serious threats, breakaways, are not allowed. They are only possible if you see to what extent you are imprisoned, how the walls are built. And that you yourself, the one who walks in the yard, are among the architects.

We can turn Marcel Proust’s idea around. Those who experience a variety of views in literature and therefore change their manners and ideas do this within the prison walls.

The task of moving from one type society and one language to another is a gigantic one. And at the same time easier than anything else. We always carry the possibility of success with us in the form of consciousness and language. We must be aware of how language works, that when we deliver language it also delivers us. If we trust our language, if we love it, the magic will happen. Some "evergreen" words and ideas will fade and blow away like autumn leaves and new ones will grow. When we move towards equality and love of each other and our environment a whole new language will insist on being born.

Chapter 7

Back to nature!

The belief in a direct link between ourselves and nature is firmly rooted among all the environmentalist authors that I have quoted. It is self-evident. But if there is a link like that, the solution to our problems is obvious: we must rediscover and strengthen that nearly lost link. Rousseau's "Back to nature" is thus repeated by many of these authors. Edberg is one writer of many who invokes "the concord with the creating forces of nature intuitively felt by primitive man". A deepened knowledge of ourselves and the universe would do the trick.

Ecology will rediscover our lost connection with Mother Nature, put us back on the right track and help us restore our "house in cosmos". This calls for "radical changes in our thinking and our attitudes, making a renewed link with the biosphere possible" (JR, MK).We have to reorganize our lives: "What is needed is a radical reconsideration of the conditions of humankind, above all to realize that we are insolubly linked to nature." (JR, AW).

When I come across statements like that, I feel an urge to protest even though I agree with the idea of concord. We are indeed linked to nature, have always been and will always be in exactly the same way. But that link is not the one the environmentalists envisage! It is language. If we follow that road instead of trying to find a non-existent link, all atrocities and environmental problems will vanish: war, animosity, hatred and the instability of our identity.

Real concord with nature can only be achieved by an understanding of the language process, an understanding that has been denied by all the theories of designation and name-giving. Language is not something that is added to reality. It is also reality. The true view of language will reveal our insoluble tie with nature: language as an instinct.

Environmentalists and scientists in this area are like persons walking around with an electric plug looking for a socket in the wall. They will never find it because the plug carries the socket with it. There are no sockets in the wall, no lost concord with nature.

A misleading pioneer in this area is Jean-Jacques Rousseau, who lived from 1712 to 1778 in Switzerland and France. His idea about our relation to nature can be summed up in one sentence: what comes from God is good, but deteriorates in our hands. That is why our real self is to be found in our innocent, natural state, when God's signature is still visible.

The misery of the world is due to our separation from this state, Rousseau contends.

In *The Origin of Inequality* he piles up evidence to show that "the larger part of our problems we have caused ourselves . . . we could have avoided them by adhering to the simple, uniform and solitary way of living that nature prescribes".

Like so many contemporary scientists and authors, he seems to think that nature (or God), not he, is behind his ideas. But nature has no voice and no philosophy. "The simple, uniform and solitary way of living" is in other words a creation of a mind formed in a certain society. And the philosophy reflects this society. Rousseau makes the same mistake as today's authors who believe that there is a socket in nature's wall waiting for a human plug.

Rousseau is constantly blaming humankind for a separation that never took place. This mistake leads to a lack of compassion on his part. He cannot see the double position of humankind: that it is a guilty instigator and simultaneously a deplorable victim.

Summing up *The Origin of Inequality* Rousseau argues that we are good-natured. His words are on fire when he shows the contrast between civilisation and the natural state:

> O you, who have never heard the voice of heaven, who think man destined only to live this little life and die in peace; you, who can resign in the midst of populous cities your fatal acquisitions, your restless spirits, your corrupt hearts and endless desires; resume, since it depends entirely on yourselves, your ancient and primitive innocence: retire to the woods, there to lose the sight and remembrance of the crimes of your contemporaries; and be not apprehensive of degrading your species, by renouncing its advances in order to renounce its vices.

This recipe for the healing of humankind was issued in 1755. Although many attempts have been made since then to implement it, the results are meagre. And no one should be surprised. In fact there never was "a natural man"! Nowhere. Not at any time. There never was "a lost and original innocence" other than as romantic, wishful thinking.

As soon as humankind appears on the horizon, language is there and puts an end to an obvious link with nature (by being this link). What Rousseau invokes on almost all pages in his writings is a phantom. Walking in the steps of this phantom he also misunderstands the origin of society in *The Social Contract* (1762). He thinks, like Hobbes, that humankind created a primitive, original society from scratch. But that is as untrue as the idea of innocence and natural man. What really happened is something quite different and unexpected: we are already trapped in a society when we create it from scratch.

In the moment when language is first used there is also a society, a rudiment, but still a form of organisation. Language moves primitive humankind away from nature and necessitates some form of organisation. This means that we not only create society. We are also created by it. To be human means irrevocably to partake in a society, without being able to do otherwise. Consequently Thomas Hobbes is wrong when he thinks that society came about to protect humankind from humankind. There never was a moment in our history when people sat down together to create a society *from scratch*. An embryo was already present at that very moment. This is the reason why our history is incoherent and bloody. We fail to see that language/society plays an important part in the creation of ourselves.

But Rousseau and Hobbes are in a way excused, as are all philosophers who lived before Wittgenstein. The decisive role of language had not yet been discovered. It is not as easy to excuse all those who live in the wake of Wittgenstein. They should know better.

What will you do if you believe in a direct link between humankind and nature? You will of course try to make it work. That is also what many people involved in environmental groups and movements have tried to do. Their theories lead them back to the soil and in some cases even to a hunter-gatherer existence that is much more difficult to revive.

I will just mention one example of the many. It derives from *Collapse*, a recently published book, where the author, on the basis of a great deal of information, foresees the breakdown of our society because of its exploitation of nature. In order to contribute to an improvement of our relation with nature, or possibly postpone the breakdown, he and some friends of his have bought a farm to create what they call an "eco-unit". It works according to the principles of ecology.

> Cultivation of fruit and vegetables, forests, pastures, streams, buildings, animals and people – all is included in a "permaculture plan" which is designed to maximize cooperation. The idea is to produce food, energy and other necessities with the lowest possible effort and in concord with nature.

Will this work? Maybe for a small, secluded group of dedicated persons who have partly detached themselves from the existing social order. But it is difficult for the whole population of the industrialized world to do the same thing: the soil would not suffice.

There are other objections as well. Will the participants of this experiment really feel that they are linked to nature, or will this be just another theoretical approach? Will they identify themselves as beings integrated in the "permaculture" or will they stand outside?

When it comes to cooperation within the group to make the "eco-unit" work, the author puts his trust in nature, this time the "human nature" of the participants. "Can we revive our instincts to live in a society of equals?" he asks himself. To me this is an attempt to lean on something that is not yet proven. Are there such natural instincts? Have they been visible in our history? The constant failure of utopians make me doubt their existence.

The members of the eco-unit are convinced that they work for a better world. But how do they know this? To me the eco-unit seems more like a modern monastery, cut off from the "real world", running away from its problems with an idealistic map in hand.

Not a single one of the environmentalists that I have come across manages to explain what is really meant by "re-establishing our link to nature". The authors do not demonstrate how this link manifests itself in production, transportation, buildings etc. Innumerable steps can of course be taken towards a resilient environment, and should be taken, but in what way do they establish a concord with nature in our hearts and strengthen our identity as human beings? In what way is our alienation diminished, our position in a "cold and unfriendly universe"?

What does it mean to us, in our hearts, in our daily behaviour, that we have 98.8 per cent of our DNA in common with chimps? This is an often used example to show our link to other beings. Do the 98.8 per cent conjure up feelings of intimacy and love of apes? Rousseau's voice reverberates in the background of these exhortations, even if the cry "return to nature" is modified. But, as I have explained, there is no nature to return to because we are already there. What remains is to discover the real link, which will solve our problems.

Chapter 8
Elitism and eco-fascism

In this book you have encountered several examples of the environmental thinking that dominated the latter part of the 20th century. The reader might think that these ideas, because of their unproductiveness, have been greatly improved or abandoned altogether since then. But contemporary books on the subject show that very little has changed.

In fact the chain of thoughts and ideas in this area seems unbroken during the past 60 years. Rousseau's voice echoes in Edberg's and other related authors and continues to echo in contemporary writing on the subject, for example in *Our Time on Earth* by Rockström and Klum. The previous gloomy dystopia in the writings of Edberg reappears in a modern form with lots of statistics and with the conviction that it is more urgent than ever to stop the deterioration of nature and work towards resilience. To do this we must abandon certain antiquated ideas that according to modern ecologists make progress slow. Our thinking deceived us. "The thought that humanity is detached from nature was never correct." (JR).

If it was never correct, how could it have seen the light of day in the first place? And even dominate the actions of humankind for thousands of years? How is it possible to cling to that idea for such a long time without discovering its falsehood? Why would it even take an effort to show that we belong to nature, if it is obvious from the start? I suspect that this thought never existed, but is conjured up by ecologists to explain why humankind pollutes nature.

"In recent times . . . we have started to discern that the assumption that humans and the biosphere are detached from each other not only is obsolete but also harmful", Rockström writes. "The consequences of adhering to false assumptions about the relation between nature and society are dramatic." Who assumed that "humans and the biosphere are detached from each other"? It seems as if many ecologists think that some kind of rational deliberation is behind our civilisation. Only this deliberation was wrong and we now have to replace it with another one.

But history was never made that way and will never be. We are driven not by reason or wisdom but by a predetermination of language that moulds our minds and our bodies.

Ecologists like the ones quoted are trying to bring humanity to reason by persuasion. They want to instigate a revolution in our thinking with a philosophy that is nearly 400 years old. They regard Enlightenment as a transition from darkness to light, a naive and unhistorical viewpoint as I see it. Now they want a similar change in the environmental area.

> Remember the Enlightenment in the middle and at the end of the 18th century. Even if it took several centuries for it to be fully accepted, it caused a "revolution of reason", that started during the renaissance and still continues, a global shift in the view of the world, pushing aside superstition for the sake of proof and reason. (JR).

"The revolution of reason" has been going on for almost 400 years, but during all that time unreason guided the actions of humankind with the belief that we are detached from nature. How is that possible? How could "proof and reason" draw that fatal conclusion?

You sometimes hear environmentalists back up their theories with the word "hubris". Earlier generations were susceptible to that. They were of course also ignorant (even though many of them lived in the era of the Enlightenment). But it seems to me that there is a considerable portion of "hubris" in the writings of modern ecologists as well. They regard themselves as representatives of reason and knowledge. The burden of enlightening the masses weighs heavily on their shoulders. It is their duty to say the right words, to push us in the direction of a stable relation with nature. And because it seems obvious that the majority does not know which way to go they appoint themselves leaders of this mission.

The sad thing is that they in taking charge of the operation obstruct it. They have no idea what truth really is: that it is something no human being can represent. No human being can possess the truth of our existence and say: This is the truth. Knowledge can of course exist in areas where craftsmanship and expertise are needed, but never when it comes to our existence as a whole. This is where many environmentalists go wrong in their ardent quest for concord.

This is also the point where the Enlightenment stumbles. It has a lot to say about the world but nothing about existence. The knowledge that the Enlightenment has given us cannot save us. We desperately need another kind, a kind that tells us why we so often act contrary to our own interest. Why do we become a menace to our own lives, our existence? The answer to this question is not be found in universities or on library shelves. No one can inform us about it. We have to live the answer to know it. Therefore we must not only examine and analyse the world around us, but also understand why we so often are victims of delusions and self-deceptions. The history of delusion is as old as humankind. St Paul writes: "The good things I want to do I do not, but the evil things that I shun I do." It is as if some alien power controlled our actions. St Paul draws the conclusion that this power is "sin" or "the devil", ideas that infected the whole of Christianity. No "sin" or "devil" is of course involved. But we are.

Environmentalists imagine themselves to be in charge. It is true. They are. Or rather could be. They regard themselves as captains of an oil tanker possible to turn around if handled in the right way. We are repeatedly told "we are in the driver's seat". Yes, we are, but only if we see how complicated it is to be get there. Drivers should know their vehicle and where they are going. Do we? We have in previous chapters examined parts of "the house in cosmos". Do we know it? The fact is that we are unable to find lasting answers to crucial questions, such as "what is language?", "what is life?", "what is a human being?", "what is consciousness?", "what is truth?" and "what is the universe?" questions that ought to be answered before all others. In what way then do we know our "house" and are "in the driver's seat" when we lack the answers to such decisive and enlightening questions?

These questions describe the totality of our predicament. They are all language predeterminations, necessary products of language and consciousness in an unequal society. These questions, and our society, fixes our heads in a certain position, like the people in Plato's cave. It is deeply tragic that what is intended to clarify obscures our thinking.

A driver who cannot answer the most important questions about his vehicle should not be driving it. He should not resign from doing so but take a step back and start the learning-process all over, realizing that driving is more complicated than he imagined. It is not impossible for us to be in the driver's seat. It is only difficult. The requirement is that we see in what way we are already determined when we make our decisions. We sometimes hear the phrase, quite often used here in Sweden: "We have to make up our minds about what kind of society we want." This is good. We have to do that. But we cannot do it without knowing the prerequisites for such a decision: how we are predetermined by language. We can compare ourselves to zebras gathered to a conference. The chairman zebra gets up on the podium and tells us: "We must decide what kind of stripes we want." All the zebras applaud.

Tim Jackson, writer of the book *Prosperity without Growth, Economics for a Finite Planet,* is another person who is convinced we are "in the driver's seat". He says that economic development is by no means inevitable – we can be in charge of it. This is of course true, but we cannot be in charge in the way he imagines us to be.

> We cannot change the ecological boundaries. We cannot change the nature of human beings. But we can create and remould our society and we already do. Its norms are our norms. Its visions are our visions. Its structures and institutions shape and are shaped by these norms and visions. This is where the transformation must take place.

It is not as simple as that. Jackson has gone to school with Rousseau and Hobbes and agrees with the idea that people have sat down together to create their society. But we are already created by a society when we create it! There is no real beginning.

The norms of our society are our norms. That is not quite true. They are not voluntary norms but rather chains that weigh down our bodies. The norms necessarily grow from the predeterminations of language which in turn grow from a society structured in a certain way. If we do not change the language by changing society our efforts are in vain. The transformation Jackson talks about will be cosmetic no matter how radical.

If we look closer to the words of Jackson these are not just *any* words. They are wrought by predeterminations of our language and society. He talks for instance about "the nature of human beings". We cannot change our nature, he says. But how do we know that such a nature exists? We take it for granted. Why? Our belief in a human nature is a great mystery, because no one has described it in full. How is it possible to talk about our nature without being able to answer the question "what is a human being?" The words of Jackson are obviously not his own (nor are his norms). They are whispered in his ears by language and society, by the way he has been brought up, by schools, class and position.

Jackson wants to show us the way to prosperity without growth. But he does so with ideas and presuppositions that have caused the problems in the first place. It becomes obvious when he compares the temptations of commercialism with the bodily inclination for sweets.

Once again he points to "human nature" – we are products of the evolution. "Cheated by our evolutionary roots, bombarded with persuasion and seduced by newfangledness we are like children in a candy shop – we know that sugar is bad for us but cannot resist the temptation." (St Paul once more!). It sounds as if he is talking to a child or an addict. And he probably believes he is. But there are no addicts in sight, only people like himself. Suddenly a political and philosophical question has become a moral one. It is only to deplore. All that is left of his rescue operation is a lame appeal to humankind: "You have to shape up!"

There is something else seething underneath the surface of Jackson's words. Can you feel the smell? I do. It does not only cling to Jackson but to everybody who ends up in the same dead end of moral appeal. It is well hidden and camouflaged by good intentions but still obvious: the contempt of ordinary people. Notice how Jackson turns to a tale of weakness when he wants to find the reasons for our predicament. So did many of the ecologists and environmentalists mentioned in this book. We cannot control ourselves. We are fooled by our own nature, our evolutionary roots. We have to fight the seduction. Jackson talks about imperfection when he ought to talk about perfection. He gives nature part of the blame when he ought to give us the full responsibility. He does so because he cannot imagine that there is such a responsibility. Everything goes wrong here as it did when we first uttered the words "human nature". And it will continue to go wrong with this idea.

Living in peace with nature takes, according to many ecologists, "new ways of thinking". "Ay, there's the rub", as Hamlet puts it. If you study books in the genre there are few signs of new thoughts. In fact, you can draw a straight line between the pioneers of the Enlightenment and today's ecologists. It is almost 400 years of the same thinking. For example, in his books Johan Rockström criticizes modern economists for their short-sightedness. They regard nature as a resource to exploit, he complains, while at the same time praising the Enlightenment. The economic model of today "does not consider the natural systems and their role in economy". (JR, AW). Partha Dasgupta, professor of Economics at Cambridge, is quoted in support.

> Nature is isolated from the economic thinking of today. Even when economists make an effort to include nature, it is treated as any other part of the capital resources destined to be exploited in the service of mankind.

Well, now. This is exactly why the Enlightenment emerged in the first place: to obtain the knowledge needed for the exploitation of nature. It is asserted now and again by Francis Bacon, the "father of science", that this is the overall purpose. That is also why his philosophy points to a strong link between human power and knowledge. The knowledge of the laws of nature gave humankind the means to "understand" nature in order to exploit it.

In their thinking ecologists, and so many other supporters of the Enlightenment, try to make 400-year-old ideas work in their care for nature. They will never succeed, if they do not deconstruct the idea of Enlightenment itself, because it is precisely that idea that has taken us to the brink of destruction. It has done so by objectifying the world around us.

We can of course objectify the world and ourselves to a certain extent. Everyone who has gone to school or been to the doctor knows that. Objectification is the shining star of science. But it is only a crutch in our dealings with existence, not a road to a final truth. If we think so we will retreat from truth instead of closing in on it in exactly the way that Sören Kierkegaard pointed out. Total objectification will lead to the destruction of existence.

> The path of objective reflection makes the subject contingent and thus existence something trivial, disappearing. Away from the subject goes the path to objective truth, and while the subject and subjectivity becomes trivial, truth will too, and this is its objective validity, because interest, as well as decision, is subjectivity . . . In its maximum this path will lead to a contradiction, because if the subject does not become totally indifferent to itself its objective endeavours are not objective enough.

Bacon and Descartes, the pioneers of the Enlightenment wanted to understand the world without the presence or participation of the subject. They really believed that it is possible to say "such is the world" and "this is the truth". Nature was God's lucid creation. Both Bacon and Descartes were eager to obliterate the harmful superstition that came with subjectivity, but their undertaking also demanded the burial of the necessary subjectivity, the one that is impossible to remove – our own presence. What grew in the footsteps of these pioneers was detrimental to humankind because science continued to disregard the subject.

The Enlightenment idea seems to be a fruitful one. After all it has survived 400 years and is still alive and kicking. Its contribution to civilisation have been fantastic: medicines, electronic equipment, all kinds of machines and what might be a more realistic view of the world around us. But there is, after all, another side to this success. When we omit the necessary subjectivity the world becomes void of ourselves. The context in which we are supposed to live takes on a strange and enigmatic look. With every step of science, this estrangement deepens. The Nobel Prize winner Steven Weinberg has remarked that the more we know about the world the stranger it seems. And the former secretary of the Swedish Academy, Peter Englund, tells us that we live in "a nature that does not care and a universe that does not notice". The Swedish physicist Ulf Danielsson infers that the universe is meaningless.

> We have a universe which is dead and meaningless from the beginning. In this meaningless universe develop, in processes themselves meaningless, living organisms and with time thinking creatures who start to reflect about questions that concern meaning. There is something rather beautiful about a picture of a meaningless cosmos that gradually comes to life and through humankind starts to reflect why this cosmos is there in the first place.

It is beyond my powers of understanding how people, without knowing the borders or the essence of the universe, still think that they can tell us what it is like. What makes human beings do this? Why step beyond the borders of knowledge into pretensions? How can anybody be sure that the universe is meaningless, and that it was so from the beginning?

Is there really no navel chord, no hot-blooded link to nature, giving us meaning and a sense of belonging? Yes, there is, as I have tried to show in the previous chapters. Language is that navel chord. We cannot shed it. We cannot doubt it. We cannot go around it or look at it from the outside. It seems as if we *are* language. It is always there, bestowing security and life on us, as is our consciousness of ourselves and the world. Without it we would not exist. This does not point to a meaningless or strange world. It *becomes meaningless* when we misunderstand consciousness and language the way that the Enlightenment does.

Just think of the words: "a nature that does not care and a universe that does not notice." Is this really true? I am absolutely sure that the person behind them cares. So do physicists, Steven Weinberg and countless others. I care too. At least my love for humankind is strong enough to make me write this book. There probably are 7 billion of us that care. So why talk about the universe as something that does not care? Are we *outside* the universe? Or are we a part of it? Why regard our own cares as something that does not exist? Or something that exists as a miracle, an anomaly that is utterly foreign to the essence of the universe?

There is something really fishy about the way science looks at the world. The idea that the world is an object has opened up breathing-space to all kinds of superstition, religion included. And it has, contrary to the intention, stripped human beings of their true identity.

The Enlightenment idea, the strategy of science, will only deepen the feeling of alienation if not paired with the insight of the necessary subjectivity. That too is rational. The answer to a scientific question like "what is truth?" Sören Kierkegaard says, "does not express the condition of the thinking subject *in his existence*, it concerns therefore a fictitious objective subject, and to confuse oneself with such is to be pulled by the nose".

To come up with new thoughts is difficult, if not impossible, if one does not understand that we are thought when we think – predetermined by language.

We are ourselves part of the problem and I am not sure that environmentalists, and all others who want to save the earth, are aware of this, at least not in the way they should be.

Almost half a century ago the idea of a well-informed and competent management of "the house in cosmos" was launched. This idea is still repeated by contemporary ecologists. But during this time, when influential people have clung to the idea, spreading it, a great number of species have become extinct, the climate is changing and the threats to our existence have grown. Maybe this unwanted development is there because the manager thinks that he knows what he does not know, and also believes that he can survey the world in its totality even though he is a part of it? Maybe the Enlightenment is counter-productive? Max Horkheimer and Theodor Adorno launch the idea in their book *Dialectics of Enlightenment* when they say that "thinking is objectifying itself into a self-generating, automatic process, mimicking the machines it produces . . . Enlightenment has given up . . . to think about thinking."

It is a very elegant way of saying that we shun responsibility by not looking at the roots and motives of our thinking. But to firm believers in the Enlightenment it is the only way, and might well be, if not coupled with the understanding of the role of a subject.

> As the most powerful species on earth we now have the possibility and the responsibility to use all our resources and all our power to enter on a grand journey, a gigantic shift to global resilience that encompasses all the societies in the world and in every step of its development considers the needs of future generations. This is the only way to assure a long term and prosperous future for humankind. We must become managers of our planet and our own future. (JR)

This is probably nothing but a magnificent delusion. If we do not see the role played by language in this journey, namely that it predetermines our thinking, we cannot manage our lives or the life of this planet in the way we want. History bears witness to this failure. It has now and again taken turns that no one really anticipated or wanted. We are not only making history – we are also its victims. This aspect is not considered in Rockström's view.

We look at animals driven by their instincts, for the most part uninterested in anything but food and reproduction and we think that we are different, that we unlike them have a free will and are able to decide our own future, what lives we want to lead and what societies to build. This is of course possible but not through believing that we are blank slates on which the writing can be controlled. The slates are full of decisive and "self-evident" writing that we have put there as if behind our own backs. Remember the words of Wittgenstein: "Language is a graft on instinctive behaviour." That links us to the animal world, giving us as much freedom as monkeys or horses have. But unlike them we can step out of our prison into another of our own choosing, if we understand the function of language/consciousness.

The belief of modern environmentalists that we can survey and manage the planet as if we were servicemen or caretakers in a factory will lead us straight into eco-fascism.

This eco-fascism and élitism was clearly demonstrated by Erik Damman, a Norwegian environmentalist from the 1970s with books like *Revolution in the Affluent Society* and *Ten Steps towards the Future.* He advocates that an educated élite should by way of information persuade ordinary people to start a mass movement, a gigantic thrust from the depth of the masses that would change the course of history. He called the movement, typically enough, *The Future in our Hands.* What happened with these magnificent plans? Was there a mass-movement? No. Nothing happened. Nothing at all. The masses did not roar, not even squeak. Very few "woke up". And that was bound to happen, and it always will, when the élite tries to influence the masses without stepping down. To be successful the movement has to turn around: the educator must be educated by those he educates, to quote Karl Marx.

As always when élitism is given free rein human beings are sorted, often regarded as either good or bad. And the duty of the good is of course to educate the bad. Rockström and Klum talk about the "broad majority of concerned citizens" who cannot afford to wait for a minority that stick their heads in the sand. Denial and reaction must not block "our common way to a brighter future." . (JR, MK).

"Our common way? It seems to me that the road is decidedly "uncommon", occupied by those who are well-informed and have the right ideas. The rest will never have a say and will simply be dragged along like a herd of sheep.

In my ears this idea of an educated avant-garde reeks with contempt of ordinary people and shows a blind eye to the history of humankind and the exploitation of human beings. It even seems as if our already frail democracy has to be restricted to save the planet.

Would it not be a better idea to do the contrary, to develop democracy instead of undermining it? Even those who believe that the earth is flat – they cannot be many – should be listened to and treated as human beings. The few cannot lead the many. That will only end in chaos as it has done in the past. Very often men cling to delusions because they are not accepted as partners in a dialogue. They become the failures of democracy.

Contempt of ordinary people point in the direction of eco-fascism, which will make miserable lives still more miserable. The first step of fascism is the usurpation of leadership.

I once met a man at a book fair who firmly believed that the earth is 6 000 years old. Many ecologists and environmentalists seem to me to have the same kind of unproven ideas. They ultimately trust in human nature to restore nature. "The will to preserve beauty on the planet resides within us. It is part of our human identity." (JR, MK).

Is it really? If such a property is part of our identity, why is it so well hidden?

The approach to history is often naive and fumbling with authors on ecology. "We have always had Mother Earth to thank for everything during our human journey . . . She has been very indulgent and our source of wealth . . . has cared for us during the last 10 000 years like a mother caring for her children, with harmony, growth and stability." (JR, MK).

This seems neither to be a picture of our history nor of nature. Maybe it is the history of dolphins, bats or crocodiles that is mistaken for that of humankind?

Ever since the first days of existence as a species humankind has suffered. We have been prey to wild animals, famine, earthquakes, tsunamis, draughts, hurricanes, decease and not to speak of everything that we inflict on ourselves, such as violence, hatred, slavery and war. What has "harmony, growth and stability" to do with these appalling experiences?

Environmentalists often regard Mother Earth as a "wise, strong and responsible parent". We have never had a parent like that. It is the belief in such a parent that creates all our problems, that makes it impossible for us to take the full responsibility. We are not born out of the womb of anyone or anything. We are, through the predetermination of language, our own children and can be nothing else. By realizing this we also discover our link to nature.

Chapter 9
Soulless robots

Among all the advocates of humankind as a product of the evolution the American philosopher Daniel Dennett is the most vehement one. In his book *Darwin's Dangerous Idea* he seems to threaten non-believers with siege and cold war.

> If you insist on teaching your children falsehoods – that the Earth is flat, that "Man" is not a product of evolution by natural selection – then you must expect, at the very least, that those of us who have freedom of speech will feel free to describe your teachings as the spreading of falsehoods, and will attempt to demonstrate this to your children at our earliest opportunity.

What he is referring to is of course the resistance to Darwinism in the American society. According to a survey published in Scientific American a few years ago, an appalling 40 per cent of Americans believe God created life 10,000 years ago. Or thereabouts.

Even so, it is still rather frightening that a person like Dennett sounds dogmatic, stepping out of reach of criticism and doubt. His words resemble those of a fanatic.

Are we products of nature? I think the answer is not as evident as it seems. All products have certain qualities that the producer provides. That is in fact what defines the word "product". It has a producer, or it would not be a product. This producer decides the qualities of the product. Should the *product* assume this task, it is no longer a product.

In other words: If evolution produced us there should be some qualities, a specific nature inscribed in us by the producer. The "production" caused by evolution could not have happened in any other way. We should be able to pinpoint a human nature, enlisting all the characteristics that we are endowed with. We should also be able to trace our steps back to our forefathers, the hunters and gatherers (because this is the point in history when our evolutionary qualities developed) and say with the Swedish anthropologist Lasse Berg: "This is the life we were made for." But are we really "made" or produced? If so where is the complete description of human nature? Where are the proofs that we are closely related to hunters and gatherers? Where are the indisputable facts that make us products?

There are signs and hints. We have a bone structure in common with all mammals, only altered here and there. And we have sight, hearing, taste and smell like many other species. But what about behavior, such as suicide, lying, deception, hate, jealousy, cunning, egotism, pride, greed, murder, rape, racism and many more aspects that are said to distinguish humankind? Is this part of our nature, something that evolution has bestowed on us? In that case court sentences should be reduced. How do you even sentence a product?

Many Darwinists and ecologists make brave attempts to find human traits in the animal world (Darwin himself tried in *The Descent of Man*), but with no real success. Their assumption is of course that evolution produces changes that will more effectively spread the genes of a certain species. But what evolutionary advantages are there in war, genocide and the murder of children? During the last 50 years humankind has been able to wipe out all life on this planet in a nuclear holocaust. Where is the evolutionary advantage in *that*?

Animals recognize individuals of the same species. We do not seem to do that. There are persons among us who divide humankind into different races, and this has led to atrocities in our history and still does. Racists do not recognize their fellow beings as such. Is there an evolutionary advantage in *that*? If all these deviations from what characterizes the rest of nature only occur among human beings, how then are we linked to other forms of life?

Dennett and all other Darwinists must answer that question. Obviously they never will because there is no "evolutionary advantage" in killing oneself. According to Darwin himself evolution could not be behind suicide. But people still kill themselves. It is even the most common cause of death among adolescents in Western Europe.

Cocksureness does not become anybody. (And you may wonder about the evolutionary advantage of *that*. Cocksureness often forebodes stumbling or failure).

But Dennett is cocksure in style throughout his book. There is not even a sign of hesitation when he quotes a poem and tells us that it sums up the totality of our lives. He does not seem to grasp that a poem is a poem, a subjective expression of the ideas of a *person*, not a statement by nature. Poems are a long way from the science that Dennett honors. But still he quotes one as the epitome of what life really is. "This is our world", he says:

> About suffering they were never wrong,
> The old Masters: how well they understood
> Its human position: how it takes place
> While someone else is eating or opening a window or just walking dully along;
> How, when the aged are reverently, passionately waiting
> For the miraculous birth, there always must be
> Children who did not specially want it to happen, skating
> On a pond at the edge of the wood:
> They never forgot
> That even the dreadful martyrdom must run its course
> Anyhow in a corner, some untidy spot
> Where the dogs go on with their doggy life and the torturer's horse
> Scratches its innocent behind on a tree.
>
> In Breughel's Icarus, for instance: how everything turns away
> Quite leisurely from the disaster; the ploughman may
> Have heard the splash, the forsaken cry,
> But for him it was not an important failure; the sun shone
> As it had to on the white legs disappearing into the green
> Water, and the expensive delicate ship that must have seen
> Something amazing, a boy falling out of the sky,
> Had somewhere to get to and sailed calmly on.
>
> (*Musée des beaux arts*, W H Auden)

What you should say about this poem is, if you regard it as an epitome: "this *seems* to be our world" or "to my knowledge this is our world". You can say nothing else simply because the poem describes how life appears to be and appearances are not lasting truths. Auden, like other authors, seems to have a glimpse of an historical instant and that is all.

To me Auden's poem says nothing about life in general, but a lot about its author, his preferences and political ideas. Neither Dennett nor Auden acknowledge this aspect of the text. But as a statement about the essence of life it is as exciting and revealing as yesterday's news. Failure to recognize one's own limitations leads to literary charlatanism. The renowned Danish critic Poul Borum pronounces a harsh judgement: "The poems of W. H. Auden seem to me not to be written by a human being but by a character in a novel by Evelyn Waugh."

There is another very strange thing about this poem: the parallel between the animal and human spheres. The dogs go on with "their doggy life" and a horse scratches its behind. Why are they there in the first place? Literary historians think that a horse painted by Breughel inspired Auden when he visited the museum in Brussels, but I do not think that is the reason. The dogs and the horse are clearly there as a background to human life, to sweep the stage for the conclusion of the poem: the same goes for human beings. We are also animals. The ploughman continues with his work and the ship although its crew saw what happened, "sailed calmly on". Dogs, horse, ploughman and sailors are in "the same boat"!

Auden is simply telling us: there is a blind, biological machinery behind us all.

"This is our world!" Dennett is very sure of this. But to me cocksureness is a sign of ignorance. He is even more convinced that we are "products of the evolution". He writes that "Mother Nature designed us to solve a certain set of problems posed by the environments in which we evolved." We are designed "products of the evolution", children of Mother Nature. Is this true? I have tried to show that there is another link to nature – language. If language is our Mother Nature it will be difficult to argue that we are "products". Then language too should be a product and science has not been able to show that it is. As I tried to show with the help of Wittgenstein, the fact is *that we not only produce language but are also produced by it.* If so the whole idea of our being "products of the evolution" stumbles and falls. Language makes it possible for us to "produce" ourselves even if the process is hidden from us in the form of predeterminations. We would have nowhere to stand and be nothing if we did not have language, which is, as I see it, the visible side of our human consciousness. We would not be conscious of a "world" or an "I" or "we", if there was no language. We would, like animals, *be* the world without being able to ask ourselves "who are we?".

There are philosophers that unlike Dennett have sensed our participation in our own creation via consciousness and language. One of them is Jean Paul Sartre. "Man becomes what he makes of himself", he says in *Existentialism is a humanism.* He goes on: "You cannot accept that any man can give an opinion about human beings as such. Existentialism liberates us from all verdicts of that kind: the existentialist never regards Man as finished, because his creation always remains." How can a *product* be unfinished, left to finish itself?

> Man is from the beginning a project that is realized through subjective life instead of being a moss, a rot or a cauliflower; nothing exists before this plan, nothing is explained in heaven (or by evolution, my comment), and man becomes to start with what he planned to become. Which is not the same as what he *wants* to become. What we generally mean by "want" is a conscious decision that comes after what you have become.

In my view these are very wise words, especially the last sentence. It is congruous with what I have said about language predetermination. If language is our Mother Nature it is obviously wrong to say that we were designed by her. In that case we are to a large extent responsible for our own design, creating ourselves as if behind our own backs. And if so the Dennett perspective must be false. To understand the creation of ourselves we have to consider Sartre's last sentence. What we become is largely predetermined. But we are of course to a large extent involved in this predetermination, having the power to change it.

If I am right, evolution is no abstract mechanism of selection as far as humankind is concerned. It is – exactly like language and consciousness – part of ourselves, of our bodies and minds. Evolution is still at work through us. This view leads to a change of our identity. We are not nature's products. We *are* nature. We can even say that we *are* the evolution.

Several things happen when we identify with the evolution. It becomes extremely difficult to say what it really is, as difficult as to say what a human being is. And we become solely responsible. We can no longer blame Mother Nature or God. In the opinion of Dennett, and many others nature is a prime subject. But if we are the evolution, so are we.

If this is true the whole system of predetermined ideas will change. We can no longer admit that we are "parachutists landing in darkness". The Enlightenment idea starts to crack. Why do we need light when we are the only light by virtue of being evolution?

This of course will seem blasphemous and sacrilegious to the scientific community. It will rise like one man and scream with Dennett: "Molecules exist independently of our experience!" Well then, scientists should try to imagine a molecule that their experience or consciousness did not discover and talk about it without language. Many linguists try to grasp the relation between language and reality. Poets write books about this relation. But they might as well talk about reality's relation to reality. It is no different with the independent molecules. They are independent in the same way that fish are independent of the water.

The question is this: Is there a reality independent of us? How is it even possible to ask that question? It will never enter the minds of people who know that they *are* the evolution and language. If we *are* reality how on earth can we be independent of ourselves?

"Mother Nature designed us to solve a certain set of problems posed by the environments in which we evolved." We became hunters and gatherers completely at home in our environment. There could have been little difference between us and the wolves in the forest or the birds in the sky. We grew out of our environment, designed for it, just like other animals. Why then did we leave that stage? How was it even possible? Why have we managed so well in environments that do not resemble the prehistoric ones?

Darwinists and evolutionary psychologists contend that we were "finished products" as hunters and gatherers. Modern man was born at that stage of our history. Does this really explain what happened? How can we be extremely adaptable and have a nature at the same time? Most biologists agree with the Swedish author Bengt Hubendick, when he says: "Biologically we are still Stone Age people adapted to a natural environment."

The statement calls for another question: How can we solve the problems facing us today, the climate change, the impoverishment of nature, if we still are designed for another kind of life? Dennett tries to answer that question by referring to the evolutionary psychologists Leda Cosmides and John Tooby. They have compared our minds to Swiss army knives.

In the varied life as hunters and gatherers we reached a high degree of versatility, comparable to that of a sophisticated Swiss army knife, they say. But is this really a useful metaphor? It seems to me that a knife like that must have an innumerable set of tools and blades, making it impossible to carry. How else could we explain the tremendous variety of human cultures, in life styles, beliefs, religions, ideas and philosophies? Did all this come out of the environment that saw the first human beings hunt? How is it possible to be designed for infinite change in a finite environment? It seems improbable. Dennett, Cosmides and Tooby try to evade the problem by saying that we are almost infinitely adaptable. But what is the point then to compare us to a finite knife? And give us traits of hunters and gatherers?

Nature gives us our identity, Dennett says, supported by the evolutionary psychologists. What identity? There is no unity on the subject. Man is good by nature, says Richard Leaky. And the Swedish anthropologist Lasse Berg agrees. No, said Thomas Hobbes, using a phrase that he had picked up from Plautus: Homo homini lupus est. Man is a wolf to man. And to Edberg we are "murder-apes" at the same time as we are a tone in nature's symphony. They are all wrong. Searching for such identities is like knocking on a bolted door.

It will never open. Not now. Not in the future. Not ever.

Dennett does not seem to understand the strength of culture or the origin and persistency of myths and beliefs. It is evident when he discusses religion. Because of the nature of the human intellect everything that is mythical and religious will be abandoned, he thinks.

> There is no future in a sacred myth. Why not? Because of our curiosity. Because . . . *we want to know why*. . . Whatever we hold precious, we cannot protect it from our curiosity, because being who we are, one of the things we deem precious is the truth. Our love of truth is surely a central element in the meaning we find in our lives.

If this is true, why did not truth conquer superstition and myths long ago? Why do they, on the contrary, seem to be more viable than ever? What happened to our "love of truth"?

Dennett makes the same mistake as scores of writers before him, like Ian Tattersall in his book *Becoming Human*. "Whatever human nature is, a fundamental aspect of it seems to be a thirst to know more about ourselves, and in particular to explain our often bizarre behavior." Nothing can be further from the truth. These words, like Dennett's, are part of the ongoing self-glorification that time after time leads humankind astray. No human being has a "natural" curiosity and an innate love of truth. All curiosity is socially and culturally determined. If this was not the case, why should a majority of scientists be interested in warfare and the construction of weapons – how to kill people in the most efficient way? On the other hand, very few, if any, are interested in research that focuses on the subject – to "think about thinking" to speak with Horkheimer and Adorno. Scientists or writers are not interested in what "consecrates" them as Pierre Bourdieu puts it, to understand their own genesis: socialization and motivation. Dennett's curiosity, as well as his search for truth, is also socially determined. He had never been curious or looked for truth in the way he does if men like Bacon, Newton, Linnaeus and Einstein had not paved the way. Science does not emerge out of nature. It is a creation by certain human beings in a certain society.

What Dennett brings to the forefront with his eulogy on truth is what Bourdieu calls a "doxa", at set of beliefs that underpins the "fields" of art, knowledge and science.

The doxa has several hidden purposes for those who believe in it. It serves to glorify artists and scientists and make them more or less immune to criticism. No one is eager to question knights on white horses in search of truth. It also serves as an ultimate motive for the endeavors of science. But most important of all: it covers up the real mechanisms, the dual economies behind cultural phenomena such as art and science. These never become visible to the public or to the artists or scientists themselves. The doxa diligently hides what Bourdieu calls a symbolic economy and a real one. The symbolic economy deals with social acceptance, status, authority, reverence, success, brilliance and other intangible but to its holders very valuable items. The real economy is intertwined with the symbolic one and bestows benefits, distinctions, wealth and appointments on those who take part in it. A vast machinery, including students, teachers, publishers, bookstores, schools, universities, prize ceremonies and much else backs up the doxa. *So while Dennett strongly criticizes myth and religion he is himself trapped in a myth,* the belief in our innate curiosity and love of truth.

You have to go through a long process of consecration to make the specific curiosity of a certain society work, Bourdieu says. The human raw material has to be altered.

> Culture, the right use of language, the good taste, was regarded – especially by those who had much of it – as general values and as

> something you owned or acquired for its own sake and from personal interest. The mastering of the world of culture and education seemed in other words to be linked to mythological ideas and an unawareness of the underlying mechanisms and the profits to come. Success in school and academic careers were explained by reference to innate talent, individual efforts or a keen interest. Artists, authors or critics often regarded themselves as unique personalities driven by love of art and nothing else. (Donald Broady summarizing Bourdieu)

Exactly the same thing can be said about scientists and their self-awareness.

Like Dennett they often refer to their love of truth and facts. Leonard Krauss does so in *A Universe from Nothing.* He praises the quest for facts. "But as I have stressed, I believe that if we are to be intellectually honest, we must make an informed choice, informed by fact, not by revelation." But this love of facts does not stop him from taking refuge in the doxa, making it the sole foundation of his own research. The scientific search for the truth about the universe "emphasizes the best about what it is to be human – our ability to imagine the vast possibilities of existence and the adventurousness to bravely explore them . . . We owe it to ourselves to draw wisdom from this experience. To do otherwise would do a disservice to all the brilliant and brave individuals who helped us reach our current state of knowledge." In his occupation Krauss has chosen "the best about what it is to be human". Congratulations!

In this grand eulogy to tradition and predecessors Krauss also gets a chance to rank himself with the brave and brilliant. You stumble across similar ideas that are designed to defend the doxa and the scientists themselves as dedicated searchers of truth. I found this in the book of a Swedish linguist: "All scientific research is based on our need to gain clarity about ourselves and the world we live in." (Bertil Malmberg)

To many ecologists human curiosity and love of truth came with the evolution.

> The adventurous life, rich in variety, developed a curious testing and trying, an appetite for the unknown that would separate man from other species. The thirst for knowledge, the will to explore was a gift from the primate to later generations. (RE).

These examples show how curiosity and love of truth are regarded as part of human nature. Nowhere do we come across the idea of them as socially developed. On the contrary the qualities are placed outside history and society, like eternal beacons of light.

Reason is regarded as a natural phenomenon, rising like a sun: "After a long-lasting groping for meaning and structure in a puzzling and complicated reality the wakening reason found its way when Enlightenment swept away superstition and religious blind faith." (UD).

In the view of this explicit adherence to and reverence of facts and truth it is very strange, as I pointed out, that scientists turn a blind eye to a lot of things that they ought to study to get the whole picture. For example: the fact is that there is always a subject present in all scientific endeavor – the scientist himself. But this fact is sadly obscured. When scientists discuss subjectivity they all gather around the unnecessary one, the beliefs and faiths, failing to see that there is also one that cannot be replaced, that has to be there as a prerequisite.

It is also a fact that the subject/scientist talks. We are trapped in a certain grammar and forced to talk about the world with that grammar. A specific human consciousness is involved in the process, the function of which science has not yet understood.

If consciousness/language is the tool of science and human understanding, it is very strange that this tool is not understood and described. How on earth could a carpenter work without knowing his tools? Consciousness should be the first thing to be understood.

Science itself is seriously impaired by the fact that it does not know everything. If you are in the situation of a parachutist landing in darkness there will always be more darkness. You might even say that Enlightenment itself stands in the way of the truthfulness of science. The whole picture is not there. And if it is not, who can say if parts of the picture, the lighted areas, are really true? To be honest: does not this make science an edifice of uncertainty?

There is yet another aspect of the doxa – to show science as a more powerful agent than it really is. This is done by stretching its knowledge and discoveries, a sleight of hand that many scientists, seemingly without being aware of it, execute with great skill.

They willingly appear on TV and radio to talk about things beyond their expertise, can barely resist the temptation to cross the boundaries of what they really know.

In her book *The Ant and the Peacock* the philosopher and biologist Helena Cronin seems to think that Darwinism explained practically everything and only crumbs are left for future scientists to pick up. Is this really so? There are lots of fundamental questions that remain unanswered as I have pointed out, questions that should have been answered before all others and would have been a long time ago if we were lovers of truth. We are not even sure of our own identity. And instead of finding the unifying "theory of everything" science moves away from it by uncovering a universe that is extremely complicated and puzzling.

Daniel Dennett has made it a habit to know more than he really does. He knows that human beings have a specific nature simply because they *must* have it. He also "knows" the structure and qualities of the first elements of life even though no one has presented them. They *must* be what he imagines them to be. This tactic becomes obvious when he tries to answer the question how reasoning, emotional and emphatic beings can develop out of blind, chemical processes. "An impersonal, unreflective, robotic, mindless little scrap of molecular machinery is the ultimate basis of all the agency, and hence meaning, and hence consciousness, in the universe." Are they really? His words make me think of some god perched high above everything telling the ignorant masses how the world is put together.

Those who oppose the idea will only get more of the same sort. Not only are we descendants of these robots, we are totally made up by what is impersonal, robotic and mindless. The thought is not a new one. It is exactly what Descartes advocated 400 years ago – minus his ideas about the soul. Descartes thought of our bodies as some kind of machines, automata separated from the soul. The soul was characterized as a "non-extended substance" outside the material world, exempt from natural laws and therefore free. The body on the other hand worked through a chain of cause and effect. The dualism of Descartes meant that body and soul were completely independent of each other, a proposition that saved him from religious criticism. The idea of the immortal soul could be maintained.

When Dennett tries to solve the problem of body and soul he simply omits one part of Descartes' dualism and keeps the other. The same strategy is used in *Consciousness explained* and that is probably why his critics have called the book "Consciousness ignored".

What he does is called reductionism. It is the favorite method of science, the altogether dominant one. Science seems to lose its footing when it cannot reduce. It is forced to limit the objects of examination to "understand" them. But in doing so parts of what belong to the desired knowledge may be omitted. This method probably has to do with the outset of Enlightenment itself. When you lack an overall understanding you have to concentrate on bits.

What if Dennett is right? Then we have to explain, not explain away, how impersonal, unreflective, robotic, mindless scraps of molecules are the basis of consciousness and empathy. This is not an easy task. And Dennett is far from coping with it.

There is also another important question to take into consideration.

How does a person who thinks he is made up of small soulless robots look at his own identity? Will there be a "chill of the soul" somewhere, generated from that thought? When I read Dennett and other authors of his kind, I detect callousness and even resignation.

I think there is something robotic in the way Dennett views the world, reminding us of the Auden poem. Animals and human beings trod along as if sleep-walking, doomed to adhere to the same individual and genetic qualities for generations. Why this view?

I suspect that the whole idea of a cold and meaningless universe is there for other reasons than Dennett is aware of. To make science possible at all you have to make the universe into an object. Then and only then can you ask yourself "what is the universe". If the universe is not an object, science has been barking up the wrong tree for 400 years. Is life an object? A human being is not. So in this universe, which is supposed to be an object, there are lots of "things" that are not objects more than in a restricted way. How is that possible? I think the scientific view of the universe results from a language predetermination. It presents a "disarmed" form of knowledge to us, one that can be used and even worshipped without stirring the structure of our societies. In all probability we have made the universe into an object to be able to exert power, to suppress and exploit one another and nature.

The choice of the word "robotic" does not seem to be an appropriate one. I do not know of any robots that copy themselves or come together in large "communities" to form complex individuals consisting of billions of cells. It would have been more appropriate to ask why something small and mindless copies itself in the first place. Authors like Edberg talk about an "urge". But that also needs further explanation. What about the thought that the first fragments of molecules, the first chemical substances that copied themselves, contained all that would follow, all the immense and astounding varieties of nature?

Could the first scrap of molecule be regarded as the Big Bang of Life? But that is no more an explanation than the dubious ideas of urges and life-forces.

Having established that we are made up of mindless robotic molecules Dennett sets out to show that these robots can create meaning. He evokes the idea of Artificial Intelligence, AI. If it is possible for us to be intelligent, it should be possible for machines too, because they are also made up of minute mindless parts. Support comes from Alan Turing, who had similar ideas: intelligent, talking machines will be constructed in a few years.

In *The Blank Slate* Steven Pinker pursues the same line of thought. Artificial Intelligence, he says, "is confirming that ordinary matter can perform feats that were supposedly performable by mental stuff alone." Just like Dennett Pinker believes that reason can arise from mindless physical processes and that "mental life can be explained in terms of information, computation, and feedback." This is an idea that dates back to Thomas Hobbes who said: "Reasoning is but reckoning." Evolutionary psychologists like Tooby and Cosmides are of the same opinion. You do not have to presuppose a "ghost in the machine" like Descartes did. It suffices to regard the brain as an advanced machine for sophisticated computation. Machines already create meaning, Pinker says. "Artificial intelligence systems have written credible short stories, composed convincing Mozart-like symphonies, drawn appealing pictures of people and landscapes and conceived clever ideas for advertisements."

Ought we to agree? Or should we ask for the programming of these machines to see what was put into them in the first place? Critical thoughts could be usefully applied here.

In recent years scepticism about the success of talking machines has grown. The problem is that the language game is an open scene where anything can happen. We are guided by the grammar, the context of our lives and social order, but machines have no such order to be guided by. "I put the umbrella in the bathtub because it was still wet" is a sentence used by the philosopher John Haugeland to show the weakness of artificial language. We immediately understand that it is not the bathtub that is wet but the umbrella, and we also immediately grasp that putting it there prevents to get water on the floor. But would a machine understand that? The professor of philosophy Martin Gustafsson at Åbo Academy says in an article that "AI scientists with great pretensions are nowadays looked upon with suspicion, especially if they aim at AGI – Artificial General Intelligence – that is believed to tie all human capabilities together". An important question is this, Gustafsson says: "What makes a goal rational? That is not a question a computer can ask itself. It is very difficult to see what it would mean that a computer has a point or an idea or grasps the meaning of things."

Critique of this kind is not noticeable in the writings of Dennett and Pinker. The latter is convinced that AI suggests and will show that "reasoning, intelligence, imagination, and creativity are forms of information processing, a well understood physical process."

I do not think that these definitions describe the human language. To me they point in the opposite direction. Dennett and Pinker seem to think that questions like "what is human language?" "how did it originate?" have been answered. To my knowledge they have not.

Every time Pinker is about to arrive at a definition, he instead mystifies it by referring to things that are not yet known (but probably will be in the near future, according to him). This is, for example, his version of the arranging principle of language:

> *Something* in the mind must be innate, if it is only mechanisms that do the learning. Something has to see a world of objects rather than a kaleidoscope of shimmering pixels. Something has to infer the content of a sentence rather than parrot back the exact wording. Something has to interpret other people's behavior as their attempts to achieve goals rather than as trajectories of jerking arms and legs.

Again we face the same body-soul problem: how subjects become objects. We are awfully close to the "ghost in the machine" here, even though its existence is denied by Pinker. We do not have to presume such a ghost or that something must be innate. Our *being* language is enough. It does the job perfectly well without the help of anything else.

Dennett, Pinker, Cosmides and Tooby look at language without understanding how it functions. They think it is based on computation and feedback. It is much more complicated than that. In the chapter about language I showed that there is a predetermination involved, that language besides being an instrument of communication also lays the foundation of our existence and our consciousness. It is impossible for us to step outside language to view it as a mere object. It is like measuring a ruler with the same ruler. But this is exactly what Dennett and Pinker are doing, as well as Chomsky, when they try to define the human language. No such definition can be made, not by anyone, and that is the first thing we should realize.

If we *are* language as well as have it, we will never get to know what language is. We will never be able to use the words: *This is language.* The construction of our human consciousness denies us all information of that kind: what we really are and what we think we are will always be two different things as long as these two perspectives do not merge, and that is a long and arduous procedure as I will show in the last chapter. The same thing with the world: what it really is and what we think it is will always be two different things as long as the model of the world is not the same as the world itself. But we can probably live sound and prosperous lives without this absolute truth, content with provisional notions.

Will there ever be an intelligent, talking machine? Of course not. Not if you want it to be as intelligent as yourself or speak a language like the human one. If such a machine is to be designed the requirement must be that it constructs itself, that it predetermines its own processes. And such a machine is unthinkable. It will never see the light of day.

The scientific idea is really a strange one, an improbable crutch, a way of finding a "disarmed" type of knowledge that leaves the foundations of society intact. At first the scientific eye transforms the world into an object, declaring that all the world consists of objects, forgetting the presence of the subject along with all presupposition and predetermination. It seems like self-deception – a person naming himself while telling others that nature named him. In doing so this person will never understand himself or the world but will remain in a squirrel's wheel, endlessly repeating the same movements.

When science tells us that it is free of predetermination we should be wary. No one is without presupposition in a world, where language uses us when we are using it.

I once exchanged letters with a scientist, a renowned one, the author of many books, and asked him if he could give me a definition of science, one that could be easily understood. He wrote back: "*There is something out there and we will find out what it is.*"

Now, this statement contains at least four predeterminations, which the scientist is completely unaware of. He really thinks that his work starts with a blank slate.

The first predetermination is "something", making the world an object. The second is "we", the self-appointment of the scientists themselves. The third is "out there", making the world alien to and independent of the researcher at the outset. The fourth states that it is possible to explain existence, a feasible scientific task to find out "what it is".

Why on earth do we not see that the show is rigged before it starts?

When Chomsky wants to explain language he does so by presupposing *another language* – innate lingual structures in the brain. It resembles the idea of the homunculus, the small figure in the brain that assembled all signals and surveyed what was going on.

It seems impossible to explain language without presupposing language. That is exactly what all the followers of the Biblical explanation do when they suppose that language is some kind of designation. Wittgenstein shows in *Philosophical Investigations* that the prerequisite of designation is – language. It is not a structure in the brain as Chomsky thinks.

I have already discussed how Chomsky compares language to a "bodily organ". The outset of every notion is always the most important, as important as the spot where a high-jumper puts his foot before he jumps. If the ground was not there the world champion could not even jump an inch. The difference is that while the high-jumper has not constructed the ground from which he leaps, Chomsky does so in comparing language to an organ.

To do this is in my view to reach a dead end. The fact is as I have already shown – and no one can deny it – that language is used to study language. Whenever was a lung used to study a lung? When was the instrument of studying a heart the same heart?

All the linguists of the world may stand up as one man affirming the value of their research, but they still cannot deny the fact that they are using language to study language. And by doing so condemning themselves to nothing but provisional, temporary results.

Chomsky's idea that you can study language as any other organ is in fact the Mother of All Mistakes, the exact point where everything that can go wrong goes wrong. This is where we are expelled from paradise (though as I will later show we never left it), the instant when we abandon our reason to be enthralled by language, seduced by ideas of a free will.

This monumental mistake is a Great Divide among scholars. They are camped on either side of a chasm like two opposing armies, the one considerably smaller than the other. In the larger camp are those who agree with Chomsky – Dennett, Pinker, Dawkins, Searle and an endless array of others, all clambering for public recognition and lucrative posts at universities and other institutions while assuring that their sole objective is truth.

In the other camp are solitaires like Bourdieu, Wittgenstein, Kierkegaard, Sartre and myself, who do not care about money, recognition or Nobel Prizes, but have concentrated our minds on clearing away all the bric-a-brac of unfounded notions that dim the light.

We are really on the verge of being ridiculous when we explain language by constructing *another* language, or study language with language. But no one ridicules those who do this, because almost everyone belongs to the Chomsky camp. They simply do not see that there is an alternative. To them there is no chasm and no other side. It does not exist. They believe science is the way to truth. If not, why were all those universities built?

Still I do not think Chomsky and Dennett are ridiculous. I just think that they are partly unaware of what they are doing. They deem it possible to answer questions like "what is language", "how did language emerge" and do not realize that their answers to such questions will be temporary. This limitation is clearly demonstrated by Dennett in *Darwin's dangerous idea*. To him language is just a tool. And "tool use is a two-way sign of intelligence; not only does it require intelligence to recognize and maintain a tool, but tool use confers intelligence on those who are lucky enough to be given the tool." Words are consequently called "mind-tools". "What happens to a human or hominid brain when it becomes equipped with words?" Dennett goes on by assuring that "when specific words . . . enter a child's brain, they find homes already partially built for them". To my ears this sounds very strange. "The words enter a child's brain". . From where? From the outside. But they also seem to be there *before* they are there, in the partially built home, the child's brain-structure. Quite amazing!

There are other oddities in this description of language. A tool is separate from its user. In what way are words separate from us enabling us to call them tools (and use them as such)? The fact is that we cannot separate ourselves from our language. If we do so, or even try, we cease to exist. If you rob the carpenter of his hammer he does not cease to exist. If you deprive the dentist of his drill he will get another one. But he is still there. What happens to us if we lose our "mind-tools", lose our human consciousness of the world and ourselves?

So Dennett is obviously wrong about language which goes for the whole Chomsky camp. It becomes even more obvious when Dennett talks about the interconnection of brains (he means people) in humanity, which gives us a superior learning momentum.

> Our brains are in effect joined together into a single cognitive system that dwarfs all others. They are joined by an innovation that has invaded our brains and no others: language. I am not making the foolish claim that all our brains are knit together by language into one gigantic mind, thinking its transnational thoughts, but, rather, that each individual human brain, thanks to its communicative links, is the beneficiary of the cognitive labors of the others in a way that gives it unprecedented powers.

I would not hesitate to call this nonsense. Even though we all have a language and a brain it does not mean that the results of brain-work and speech are worth the effort.

Others cannot be responsible for my words. I have to find the truth myself and cannot rely on others to do that for me. All my energy should therefore be directed towards this goal.

This is clear from what I have just demonstrated. The side of Wittgenstein has nothing in common with the side of Chomsky. Nothing at all. There is simply no communication here. So in what way are these two parties "beneficiary of the cognitive labors of the others"?

All the efforts Wittgenstein has made and all the seeds he has sown have never fallen into the soft soil of the other side but has been lying there to no use. And still Wittgenstein tried to communicate all his life. *On Certainty* was written weeks before his death.

Communication and the efforts of others do not guarantee anything whatsoever. All this seemingly beneficial labor might be a cloud of dust. We simply do not have the means to decide the relevance. The powers that Dennett talks about might well be fool's gold.

In Dennett's mind language also seems to be the origin of science. "Once we have language – a bountiful kit of mind-tools – we can use these tools in the structure of deliberate, foresightful generate-and-test known as *science.* All the other varieties of generate-and-test are willy-nilly." It seems to me that Dennett regards science as something that has sprung out of nature without being touched by human hands, an edifice beyond criticism.

Science to Dennett is the "natural way" of looking at things. But science does not grow from nature like a tree. It has a history and a starting point of ideas. It needs predetermined notions to get started and to maintain its momentum. It needed Francis Bacon.

Dennett's attempts to show how meaning is derived from lack of meaning are not convincing. To believe that there is a mental grammar in the brain that in some way directs our notions of the world is like believing that the head of a hammer contains blueprints of everything you can construct with it. The function of consciousness and thus of language can only be described as a paradox – "it is not what it is" – as Sartre has put it. With those words in mind it is both entertaining and saddening to come across "explanations" of language and consciousness as if they were objects. How do you explain something *that is not what it is*?

The words of Jean Paul Sartre in *Being and Nothingness* have been the guiding star of my thoughts throughout this book. They vividly denounce Dennett's naturalism.

> Consciousness is, because it is producing itself, which is due to the fact that its being is consciousness of being. This means however that the doing (le faire) maintains the being; consciousness must be its own being, it is never maintained by being; consciousness maintains being in the heart of subjectivity, which consequently means that it is inhabited by being but it is not being; *consciousness is not what it is.*

It took me a long time, almost 20 years, to understand what Sartre says here. And I think it is difficult to grasp the meaning without the philosophy of Kierkegaard and Wittgenstein. But now these words are as clear to me as the blue sky on a summer's day. They simply say that consciousness and thus human language maintain themselves, *even though* they are maintained by the world around us. Consciousness produces itself even though it does not.

This makes the function of language/consciousness extremely difficult to understand. And in no way should we compare it to a tool. Language is something completely different: the builder of worlds that are and at the same time are not, worlds that are real and solid but at the same time lack these properties. This function of language and consciousness is what makes the world real. We do not live in a dream like some philosophers have suggested. And our knowledge of the world cannot be summed up by "will and representation" as Schopenhauer thought. Our representations *are all there is without being all there is.*

The Darwinian approach to language, interpreted by Dennett, is clearly a failure. The same thing happens when Dennett discusses the impact of evolution on religion.

He is convinced that Darwinism sooner or later will do away with religion. In the future we will only encounter religious ideas at our museums where we will be informed of the strange habits of people long ago. Darwin's idea "threatens in the long run to be just as toxic to these (religious) memes as civilization in general has been toxic to the large wild mammals." "We must face the fact that, just as there were times when tigers would not have been viable, times are coming when they will no longer be viable, except in zoos and other preserves, and the same is true of many of the treasures in our cultural heritage."

This is of course also a misunderstanding parallel to the misapprehension of language. Dennett misjudges the strength of culture and language. He does not acknowledge the ability of language to create identity, does not realize that we *are* language as well as have it. People will, in other words, and they have done so, defend religion with their bodies, because they experience religion as vital to themselves and will do so in every unequal society.

In his fierce attack on fanaticism Dennett lists all creeds where fanatics and fundamentalists may be found – and should be caged like animals: "Protestantism, Catholicism, Judaism, Islam, Hinduism, Buddhism, as well as countless smaller infections". But there is one creed of fundamentalists missing. That is the science community, where cocksureness and fundamentalism grow from what is deemed to be "undisputable facts". It is an "undisputable fact that we human beings are products of evolution" Dennett proclaims. This is fundamentalism and like all fanaticism not true. We are not products of evolution, because we *are* at the same time evolution, responsible for ourselves and the world.

I am convinced that humankind with language has received a gift of a magnitude that is still unrecognized. This wonderful device has not been fully understood by those who regard humankind and language as products of nature. They fail to see the full potential of language – that it is a factor in creating our existence. Omitting that, they have not even understood what evolution really is; that we are the heirs of the evolutionary process, its chosen champions. Those who position humankind in a Darwinian scheme without understanding the real function of language reduce our stature and make it impossible for us to reach a stable identity. If we accept the view of Dennett, which is the same as accepting the dominating view of science, it will be very difficult to solve the serious problems that face us.

All of us who use the word "creation" or "product" about human beings misunderstand our identity. We belittle ourselves as primary subjects acting in the world. And we do this through predetermination because we live in a society driven by inequality. To really understand ourselves we have to replace the direct link between us and nature with the direct link between us and the nature of language. This makes us fully autonomous.

These misunderstandings have a source in the writings of Charles Darwin.

> As man is a social animal, it is almost certain that he would inherit a tendency to be faithful to his comrades, and obedient to the leader of his tribe; for these qualities are common to most social animals. He would consequently possess some capacity for self-command. He would from an inherited tendency be willing to defend, in concert with others, his fellow-men; and would be ready to aid them in any way, which did not too greatly interfere with his own welfare or his own strong desires. (*The Descent of Man*)

Darwin here romanticizes humankind like Rousseau did (remember that he just a few chapters ago did the opposite when siding with the apes). His predetermined starting-point (we are social animals) – that answers most of his questions before they are asked – sets him off in the wrong direction. He presupposes that humankind has got its qualities from nature and evolution, which means that we are only partly responsible for ourselves.

He leans heavily on his discovery of evolution. Expressions like "inherited tendencies" are everywhere in his text – you cannot even begin to count them. Also notice the way in which the will to aid our fellow-men is restricted. Do we really behave like that? To my knowledge the will to help others vary between total unwillingness and risking one's life.

What you really meet in this text is not an attempt to find the link between humankind and nature but the confessions of a person estranged from himself. Darwin fails in the search of a human identity (which would make him content) and is thus forced to concoct a provisional one, propped up by immense efforts and all the various specimen collected from nature. The self-alienation of Darwin must have been considerable, judging from the enormous energy he put into discovering the natural world. His endeavour makes me travesty some words in the Bible: what is the point of knowing the whole world if you do not know yourself.

Dennett makes the same mistake as Darwin, believing that human beings are "products of the evolution". Neither we nor the rest of creation are *products* of the evolution. Algae, plants and animals were and *are* the evolution, even though there is an evolutionary process involved in their existence, just like in ours. Dennett and Darwin, and many others, regard themselves and us as bystanders eyeing the wonders of the natural world. But that is not our primary role or position. We are participants. We play a crucial part in the drama of the development of life. It is very strange that we in such a context have arrived at the conclusion that we are products of the evolution, *children* of Mother Nature and thus secondary.

How do you pinpoint someone who is always in motion? To define him you must make him stand still. But then you strip him of his foremost quality. That is how the evolution is treated and that is also why nobody can answer the question "what is the evolution?" As soon as you deliver the answer, it will prove wrong. The object of your studies is on the move.

Our language cannot express any definite truths for the same reason. It is part of life, part of ourselves. We *are* language and that makes it impossible to answer the question “what is language?” On the other hand, it is only through language that we can rid ourselves of the alienation that is so obvious in the writings of Darwin, Dennett and many others.

To be the word is not a privilege of gods or Jesus-figures. It is the birthright of human beings everywhere, a prerogative closer to us than anything else.

Chapter 10
Political biology

"The condition of the earth is critical." This is what you hear from ecologists and environmentalists all over the world (here from the biologist Staffan Ulfstrand).

It is therefore of the utmost importance to change direction. This requires, according to Ulfstrand, a cultural revolution that will bring "new norms, new attitudes, new ways of living and new goals for humanity – a future quite different from our past".

In view of past mistakes, it sounds like a wise recommendation. Change is not only necessary. It is urgent. Many areas of life are threatened. According to Ulfstrand and many other biologists, there is a huge obstacle, however: "Human nature is a mixture of good and evil. It has been that way since the beginning of time and will remain the same. We cannot do anything about it." In other words: we can only be partly successful. Nature rules.

It is very fortunate that this is political biology. If it were not, we would never overcome the crisis that becomes more threatening every day. The evil part of our nature would then always be there to obstruct and destroy our good intentions and make the future a sordid copy of the past. There are of course evil acts in history. No one can deny that. But they cannot be explained by referring to a "human nature", because no one has shown, in a scientific manner that such a nature exists. Many biologists take it for granted, as Ulfstrand does. But taking something for granted is not science. As long as biologists have not demonstrated the nuts and bolts of human nature, there is no weight to arguments based on it.

Ideas like Ulfstrand's are counterproductive. On the one hand he contends that "we are masters of our destiny" and on the other "we have to put up with ourselves the way we are". How can we map out our destiny if something else – "the way we are" – is standing in our way, directing our thoughts and actions? These contradicting ideas make biologists pirouette on the spot like well-trained circus-horses. Nothing positive will come out of such ideas.

Scientists know that the idea of a human nature is a feeble one and have therefore tried to bolster it. Donald E Brown, an American anthropologist with ties to the University of California has made a monumental effort to solve the problem once and for all.

He has compiled an extensive list of *Human Universals* (the title of his book). But even from the outset this book is, in my opinion, a complete failure. To be able to put anything at all on the list we have to survey *all of history* and *all cultures.* Such an immense undertaking lies beyond our means. And even if we should manage, this godlike knowledge would tell us nothing about our nature, because history is not finished and cannot be summed up.

One of the items on the list is “belief in supernatural/religion”. Such beliefs have indeed accompanied humankind throughout history. But that is no guarantee of universality. The same thing can be said about “beliefs about fortune and misfortune”. Even though such beliefs are widespread, especially in the American and Western societies, there is nothing universal about them. They are part of a culture and not of an eternal destiny.

“Childhood fear of strangers” is another debatable item on the list as is “conflict”, “copulation normally conducted in privacy”, “contempt”, “discrepancies between speech, thought, and action” (there is at least one person who displayed no such discrepancies and thus claimed he was the truth), “division of labour by sex”, “economic inequalities”, “envy”, (there are probably many human beings who in this respect lack a human nature), “facial expression of disgust”, “females do more direct child-care”, “gossip”, “in-group distinguished from out-group”, “insulting”, “judging others”, “language employed to misinform or mislead”, “language, prestige from proficient use of”, “magic”, “males more aggressive”, “male and female seen as having different natures”, “males dominate public realm”, “manipulate social relations”, “prestige inequalities”, “property”, “rape”, (how on earth do you prove that rape is a part of human nature? Because it is frequent in many *societies*?) “sexual jealousy”, “sexual modesty”, “special speech for special occasions”, “statuses”, “tabooed foods”, “territoriality”, “weapons”, “dominance/submission”, (suggests that there will never be, or have been, the possibility of equality in any society), “fear of death”, “husband older than wife on average”, “males travel greater distances over life-time”, “pride”, “manipulation of self-image”, “shame”, “thumb sucking”, “violence”.

It seems to me that professor Brown in order to compile his list has been looking out of the window at his university to catalogue the American way of life. And what else can a person do if he wants to describe human nature? If there is a human nature, which I doubt, it should be possible to enlist human Universals, behaviour that keeps recurring. But there is nothing of the sort in sight. Brown's list is just a catalogue of what human beings *seem* to be, or what has been manifested as human practice in certain societies. As we will see later on the identity of humankind, of the self, is much more complex than listings in a catalogue.

Science cannot start from a blank slate, nor can religion. If you ask yourself "what is a human being?" it seems that you start at zero. But you do not. "What" is already leading the way. The question is pregnant with the answer just like the question "Is there a God?"

Darwin's answers are no exceptions. Like all scientists before and after him he was enmeshed in a scientific tradition, asking questions in a scientific way. When he seemed to have perfected his theory of evolution, the question remained: how does humankind fit into the scheme? The answer could not deviate from what he had already discovered or the theory would be jeopardized. So Darwin made a great effort to define humankind's place in nature in *The Descent of Man.* The ground was prepared in the *Origin of Species*, where he says that "light will be thrown on the origin of man and his history". "Man must be included with other organic beings in any general conclusion respecting his manner of appearance on earth." In the introduction to *The Descent of Man*, where Darwin also outlined the goal of the book, he wrote: "The sole object of this work is to consider, firstly, whether man, like every other species, is descended from some pre-existing form; secondly, the manner of his development; and thirdly, the value of the differences between the so-called races of man."

What Darwin has to do in order to reach his threefold goal is to show the similarities between human beings and other animals. When it comes to mental faculties, no one can deny that there is a great gap between humankind and the higher animals. Even Darwin acknowledged that. But the difference is just one of quality. "My object . . . is to show that there is no fundamental difference between man and the higher mammals in their mental faculties", he says and goes on with a very strange and illogical conclusion: "as man possesses the same senses as the lower animals, his fundamental intuitions must be the same". The idea is that if you have all senses in common you must behave in a similar way. He seems not quite sure of what he just said and takes a step back in the next sentences. "But man, perhaps, has somewhat fewer instincts than those possessed by the animals which come next to him in the series." He reinforces the idea of similarities by saying that "most of the complex emotions are common to the higher animals and ourselves" and goes on to prove it.

Just like humans, animals show wonder, curiosity, imitation, attention, memory, love, temper, imagination, reason, self-consciousness, ability to abstract, language, a sense of beauty and other such capacities. The list is substantial and overwhelming.

> Dogs show what may be fairly called a sense of humour, as distinct from mere play; if a bit of stick or other such object be thrown to one, he will often carry it away for a short distance; and then squatting down with it on the ground close before him, will wait until his master comes quite close to take it away. The dog will then seize it and rush away in triumph, repeating the same manoeuvre, and evidently enjoying the practical joke.

This is an exact description of what dogs did during the 19th century and *still do*. For thousands of years dogs have had an unchangeable "sense of humour". I think Darwin is using the wrong word here. If dogs had humour they would surely not repeat themselves.

It has been said that Darwin loved animals and revered the fantastic development that he managed to describe in *The Origin of Species*. In my opinion, there is very little of this reverence in his later book. In his urgent quest to prove his theories he does not seem to respect the integrity of animals. He adorns them with all kinds of human emotions and capacities like that of humour. The approach smacks of anthropomorphism.

This strategy is repeated by Darwin's followers. The similarities between us and other living creatures must be accounted for, or they would not be our ancestors. In Sweden a modern replica of *The Descent of Man* is *The Heritage of Virtue*, written by psychiatry professor Nils Uddenberg. He does what many Darwinists do. To convince us of our heritage he stretches both ends making humans more ape-like and apes more human-like.

What is the first thing that you must remove from humankind if you want to compare us to apes? Unpredictability. We are very often unpredictable in our actions and even to ourselves. Many biologists and ecologists realize that this is a crucial point. Apes have to be made more unpredictable and human beings less if the equation is to even up.

We are all conscious of the unpredictable side of others and ourselves. Nothing has to be the way it seems to be. The greatest feeling of all, the love of another human being, is expressed in uncountable ways. Love does not even have to surface, but be unknown to everybody, shining like an inner light making the world meaningful. The characteristics do not have to be those of novels or films. It seems that there is always something left to describe, an opening for new expressions. The same is true of other emotions. If somebody hates me it does not mean I have to react in a similar way. I might turn the other cheek or just shrug my shoulders. There are innumerable examples of similar behaviour.

Emotions do not even have to be emotions, but can easily be mimicked.

How does Uddenberg handle the issue? He predominantly describes human feelings and reactions as biologically determined. “Take a couple of friends who have cooperated in a business or at a science department. When one of them is letting the other down the fury of the deceived person cannot be stopped.” Is this really what *always* will happen?

On the contrary, the person in this example can react in numerous ways, with rage, with surprise, wonder, sorrow, indulgence, forgiveness and even compassion. All depends on the interpretation of the act. The deception does not even have to be deception.

When I lived in Canada in the 70s, my wife worked together with a person who seemed nice and trustworthy. Both were psychologists with the aim to develop an instrument that could measure retarded students’ adaptation to industrial work. Of the two my wife obviously came up with the most important ideas and the work progressed. But when the report was handed to the management there was only one name on it: that of her partner. She asked him about it. Was there an oversight? Had he forgotten to put her name on the report? His explanation was utterly astonishing to both of us. You must use your elbows to get anywhere, he admonished her. And the sharper they are the better. He even blamed her: She should have gone to the management before he did to show them how clever she was. Biologists like Uddenberg would never have guessed our reaction. We felt sorry for the man, for his upbringing, for the society he lived in, for his egotism and his future life. In our view he was not a guiding-star of future psychologists but a person in need of help.

Human reactions are like this, difficult to predict. But biologists wholeheartedly trust in biology. "Human beings have like other animals a biologic tendency to react with anger in certain circumstances." (NU). "The theory of evolution can explain what situations cause anger." (NU). What circumstances and situations? Biologists are often reluctant to tell us and when they try examples become ridiculous. "He hit me so I hit him back" (NU).

Ideas like this may well contribute to the high level of violence in our societies and to apologetic behaviour. "I did nothing wrong. I just acted according to my nature."

Leaning on these observations Uddenberg argues that we did not invent the exchange of favours even if reciprocity is very common in our relations. It is therefore in our nature to feel "a duty to return what others have given us by paying them back with the same value". The ape-inherited reciprocity initiates dependency in exchange of values. "The person who has been treated with benevolence feels like a debtor and regards the benefactor as a creditor."

This is also, he argues, the foundation of the golden rule. There is a debt behind it. According to the Bible, "we are told to be benevolent to all people no matter if they have been so to us or not. But behind the benevolence there is an expectation of the favour returned."

In my opinion the golden rule is thwarted by this addition referring to evolutionary roots. The altruism imbedded in the golden rule becomes selfishness. We expect something back. I wonder if the biologist version of the golden rule has anything to do with it. Is it really based on reciprocity? I do not think so. On the contrary, I think reciprocity will ruin it. The golden rule exists because we love our fellow men, not because we love ourselves.

When the golden rule is applied, there is an absence of reciprocity. Primo Levi tells of one instant in *Is this a human being*? And Alexander Solzhenitsyn has examples from Gulag. If you are a strict Darwinist you might expect that individuals who are always hungry, frozen and exhausted would take every opportunity to grab food from others using all kinds of trickery and deceit. There are of course such acts. But there are also examples of the opposite, where people have shared their rations even though they jeopardize their own lives in doing so. They obviously do not expect anything in return as the recipient lacks the means. You might argue that there still might be some selfishness involved: the satisfaction of being good. But does this really apply to persons depicted by Levi and Solzhenitsyn?

Unpredictability is abundant in human relations. But to biologists this fact is of minor concern. They are convinced that we have a nature that translates into actions.

> When evolutionary biologists place humankind on a level with animals, they do this because the same biological laws that gave other living organisms their characteristics also have shaped our own. By studying the forms of life close to us we can apprehend why we have become what we are. (NU)

Many philosophers object to this simplified way of grouping human beings with "other living organisms". One of them is Jean Paul Sartre, whose view on the formation of human beings is quite different from that of Uddenberg and many biologists.

Sartre disengages us from nature. Biologists and ecologists do the opposite. In order to link us closely to nature they try to find the similarities between man and beast. But what is human nature? There is no reliable account to be found anywhere to make the comparison possible. If science does not have the answer, literature surely must. It has been around for thousands of years. This is also where Uddenberg finds support, just like Dennett did. And he does so in a book from the 16th century by Nicolò Machiavelli, *The Prince.*

He is helped along by science in his pursuit. Primatologists like Frans de Waal point to a "Machiavellian intelligence" among the larger apes. It seems as if their observations prove the closeness of our relationship. But I think the primatologists forget something, just like Dennett did with his idea of an archetypal poem. *The Prince* by Machiavelli is *literature* and as such a very subjective account of human life. It has nothing to do with humankind as such but offers advice to princes who want to gain power or maintain their position. It should be obvious that princes can no more represent humankind than can cannibals or paedophiles.

What characterizes Machiavellian intelligence that show the link between us and the larger apes? Both species intrigue, fawn, build hierarchies, humiliate each other, bestow favours on close kin, are familiar with the idea of reciprocity and much else. The conclusion is obvious: "Politics is older than humankind . . . Friendship and the greed for power, revenge and reconciliation are phenomena with roots far beyond the human sphere." (NU).

The Machiavellian chapter ends with even more general claims:

> All human culture is partly determined by our evolutionary history. Our consciousness as well as our empathy can be understood according to the conditions of one who lives in a group of cunning apes. When the advice Machiavelli gives his prince reminds you of the tricks that are used in the struggle for power among the apes at Arnhem Zoo it is more than a laughing matter. The princes of Machiavelli behaved as they did because they were descended from the same social being that also was the ancestor of the big apes. (NU).

According to biologists it is obvious that we are kept on a leash by nature, even though the length varies. To some biologists it is extensive as is the free will, to others rather short.

On the one hand we have a human nature that influences our actions. On the other hand we can surmount the call of nature and act as if oblivious of it. But only to a certain extent, because "we were born selfish" as Richard Dawkins puts it in *The Selfish Gene.*

Uddenberg pursues the same line of thought. We can adjust the leash but only to a certain extent. He even fears that the leash is more coherent with the dark side of human behaviour than with our positive qualities. "The ambition to revenge – the desire to punish the violator of the contract – is an inevitable complement of reciprocity." As I have tried to show – very little is inevitable when it comes to human behaviour. But biologists who are eagerly trying to link us to nature, turn a deaf ear to such objections. Instead they often tighten the leash with an array of statements: "Violence is part of our nature". "Aggressiveness can be regarded as a biologically founded behaviour." "The desire to revenge is a fundamental human inclination." "To retaliate upon an injustice seems to be one of our most fundamental impulses, as irresistible as hunger or sexual lust." (NU). Does this strange repetition, dwelling exclusively on actions spurned by hatred, revenge and aggression, really point to fundamental qualities of humankind? It seems more like the talk of a disillusioned and bitter person.

Uddenberg even lets xenophobia be a product of evolution telling us "the tendency to make a difference between 'we' and 'them' is deeply rooted in human nature."

Reading this I cannot help shivering with disgust. This is pure political biology, subjective ideas masquerading as scientific facts in order to strengthen inequality.

Trying to prove his point Uddenberg resorts to the same idea that I paid attention to in the opening of this chapter, the existence of human universals. "There is a core of human behaviour and notions that are repeated from society to society and this core is probably handed down by evolution." The idea is repeated in the text like a mantra. "The emotions of human beings have a given form: love and loyalty is like jealousy and revenge universal human passions." "All these universals indicate a human nature."

> Beyond human diversity there are human universals: patterns that are repeated from place to place and in different ages. We recognize ourselves in the texts of Homer, in Chinese love poetry from the 14th century or in the alliterated verses of the Edda poetry. We are not only cultural beings – we also have a nature.

Is he really talking about genuine universals? Or might there not be other explanations to the historical uniformity, for example that societies where these passions appear fundamentally resemble each other: they all thrive on inequality. In my opinion a truly scientific author would have written: "There *seem to be* certain universals."

In a later book, *Ideas of Life*, Uddenberg clings to the same notion of universals and the importance of biology. "The person who wants to understand why certain types of societies have a tendency to repeat themselves must study the fundamentals of evolutionary biology."

There is little hope for humankind in this. We seem to be stuck in a treadmill of endless repetition. In the face of such a grim destiny it is consoling to know that the idea of universals in history and in human beings is a backbone of political conservatism.

It is designed to resist social change and preserve privileges and injustice.

The notion of a "human nature" is coupled to the idea that we are products of evolution. Uddenberg and other biologists only repeat what Darwin said in 1859. And they do it not because of the truth of the matter, but as members of a Darwinian tradition.

A "human nature" has been given to us by evolution, Darwin says. If so, consequences will follow. What about the roles in society? The differences between sexes? Are they too determined, at least partly, by nature? Darwin's answer is in the affirmative. In order to prove his point he drags a whole zoo into the discussion. "No one disputes that the bull differs in disposition from the cow, the wild-boar from the sow, the stallion from the mare . . . and the males of the larger apes from the females." Darwin is convinced that men and women are different because the sexes of animals are. Women are thus by nature more caring and less selfish than are men, which make women resemble savages. Men, on the other hand, are rivals to a greater extent. They like competition, have a higher degree of ambition, which makes them more selfish than women (and less related to savages). They lack the intuition, tenderness and rapid perception of women, who also are better imitators, qualities that are "characteristic of the lower races, and therefore of a past and lower state of civilisation."

When Darwin tries to demonstrate the "bull-and-cow" theory in humankind he does so by listing the achievements of men and women in society.

> The chief distinction in the intellectual powers of the two sexes is shown by man's attaining to a higher eminence, in whatever he takes up, than can woman – whether requiring deep thought, reason, or imagination, or merely the use of the senses and hands. If two lists were made of the most eminent men and women in poetry, painting, sculpture, music . . . history, science, and philosophy, with half-a-dozen names under each subject, the two lists would not bear comparison. We may also infer . . . that if men are capable of a decided pre-eminence over women in many subjects, the average of mental power in man must be above that of woman.

I wonder what Darwin would have said about the intellectual capacity of boys and girls in Swedish schools, where the girls vastly outperform the boys – even immigrant girls outperform native Swedish boys. It should be obvious to everybody that Darwin's comparison is nothing but unjust, clumsy and ill-concocted. One could say that Darwin falls into a trap. He starts off with the notion that differences are natural and that they therefore are expressed in social life. He thereby omits all the real causes of these differences: tradition, opportunities, patriarchy, child-bearing, home-making, schooling, professions, social order etc.

He simply does not see that his conclusions are biased, that he disguises what is really the cause of inequality and differences under a veil of putative science.

Like somnambulists biologists walk in the footsteps of their master:

"The social roles of men and women are partly an expression of their biologically determined nature." (NU). One wonders about the "partly" bit. What part is that?

In accordance with this, certain male characteristics are unavoidable. Uddenberg tells us the story of an executive at a Swedish export company in the US who had demanded 'sexual favours' of a female employee. "Without excusing this executive one must establish the fact that he behaved just as you expect a successful primate male to behave."

There seems to be some kind of predetermined pattern of behaviour in "successful primate males". How do less successful or even unsuccessful primate males behave? Do they not have the same nature, or are "successful males" some kind of new species?

"Men have . . . a biologically founded will to make social careers." It is convenient to resort to biology to preserve privileges: it makes it an act of nature that around 90 per cent of Swedish university professors are men. Then no one will look for the real causes.

"The background of the traditional distribution of power between the sexes is found in our biological history." This is really good news (to men). Evolution is on our side!

Biology even dictates wage policy. I wonder if the negotiators have discovered this powerful ally? "The unequal distribution of power and income in society is a modern reflection of the social reality of the ape-flock." (NU). We are told that there is a biological reason for the differences between male and female salaries! Uddenberg realizes that he might be walking on thin ice and modifies his statement: "Even if a plausible evolutionary explanation is possible as to why men very often have higher salaries than women – and such an explanation is possible – it is not the same as saying that this is how it *should* be."

In *The Heritage of Virtue*, just as in *The Descent of Man*, we encounter prejudice against men and women, disguised as science. Beneath the scientific veneer something is moving like a beast shunning the light of day: a conservative, bourgeois consciousness.

The same political consciousness holds humankind to be partly incorrigible and blames that on the evolution. Darwin himself does so to a certain extent. Edberg, Dennett and other ecologists that I have studied agree with him, even if they also stress our adaptability.

Richard Dawkins emphasizes that humankind to a certain extent is incorrigible. In *The Selfish Gene* he issues a warning: do not believe too firmly in social change.

> Be warned that if you wish, as I do, to build a society in which individuals cooperate generously and unselfishly towards a common good, you can expect little help from biological nature. Let us try to teach generosity and altruism, because we are born selfish. Let us understand what our own selfish genes are up to, because we may then at least have the chance to upset their designs, something that no other species has ever aspired to.

This seems to me to be another example of the split human being demonstrated by ecologists and to a certain extent by Darwin himself. Generosity and altruism have to be taught, harnessing a deeply rooted natural selfishness that will reappear with every new generation. It is like imagining a fly steering a 16,000 horsepower oil tanker.

To make his theories seem plausible Dawkins emphasizes that even babies are selfish. And they must be, if the selfish-gene theory is correct. Generosity is secondary.

"Selfish greed seems to characterise much of child behaviour", Dawkins says. Is greed really the proper word here? Is a baby sucking the milk of its mother greedy?

Another question also has to be answered. Why is it possible for us to teach generosity and altruism, if we were born selfish? How and where does that urge originate?

The core of Darwinism is that *all* human beings have a nature. And if so, *all* human beings have to fight that nature in the way Dawkins describes. But all of us do not. There are persons, even in our competitive, individualistic society that do not fight this nature at all. To them it is not even an effort to be unselfish and generous. Where is their human nature?

The subject in Dawkins' notorious book is not a human being. It is the gene. We are only vehicles built to transport genes from one generation to another, Dawkins argues. I have previously tried to show that super-subjects like God or Nature exist because we are forced to behave like secondary subjects in all unequal societies. We have to install super-subjects in our view of the world to make it seem rational. The idea of a God or Nature that have created or produced humankind is therefore a political statement, not a biological or religious one.

The same thing can be said of the gene. It is a supreme subject ousting the only subject there is, the human being. The reasons are social. Dawkins is political when he thinks he is scientific. We should remind ourselves of the words of Karl Marx, when subjects like Nature, God or The Gene usurp our birthrights: "The highest being of man is man himself."

In *The Heritage of Virtue* one chapter is titled "Utopias and The Incorrigible Human Being". The theory that "the evil of mankind is caused by an evil society" annoys the author. To prove its falsity he points to the catastrophic and tragic attempts to create ideal societies. Their failures are due to biology. Human nature is difficult to change.

He criticizes Emile Durkheim and Margaret Mead for their socially based theories. It is utterly wrong he says that "culture only can be understood from culture" (Durkheim) and that "the personality types of the two sexes are social products". (Mead).

Modern science, especially biology, has made such theories obsolete. Psychologists, sociologists, anthropologists and other scientists are abandoning Durkheim and Mead in favour of biology, the only science that reveals "the essence of humankind".

Even if social change is rapid, the characteristics of humankind were shaped by conditions prevalent during the hunter and gatherer era and cannot be transformed at will.

We cannot erase all our undesired qualities with pedagogy and good intentions but have to live with our biologically determined imperfections. The failure of Utopians ought to teach us one thing, Uddenberg emphasizes: we are not a wax in our own hands. "Humankind cannot create itself at will, there are certain biological patterns that we cannot disregard."

Could there not be another reason why Utopians fail? I would argue that they fail because they regard what is present, our contemporary time, as insufficient. Utopians locate the "ideal state" in the future, just as religious people locate paradise in a life after death. This is the fundamental mistake, because our present existence supplies us with all the means to solve all problems. We just have to understand how language predetermination works. We need not look for answers in the future or search for them in heaven (or in the evolution). *Our true identity is always with us* though not discernible in a solely scientific way. Utopia is escapism as well as dystopia. And that so many take refuge in one or the other is utterly tragic. Utopians are like starving men who turn away from a rich table to search for it in the sky.

Ignorance of language function is behind the failure of Utopians. If you try to change society without changing the language you will never succeed. The old society will hibernate inside the new one and wait for a ray of stimulation to show its hidden claws.

Evolutionists, biologists and religious people, believing that they were created or produced, do not seem to understand the importance of the subject, of their own lives. For that reason their attempts, if there are any, to show humanity the way out of a threatening ecological breakdown is counterproductive. They bolt the door in trying to open it.

A future breakdown of our means to breathe, feed and live will not be caused by the constancy of our nature or our lack of knowledge. It has nothing to do with hubris, selfishness, insufficient love of our neighbours, or any so called human deficiency. It solely rests on our inability or unwillingness to acknowledge that our "ecological system" is different from that of all other species and all other natural systems. Language/consciousness is our true breeding ground, our natural system that will reconcile us with nature.

Chapter 11

The function of consciousness

From the first moment of our history as human beings we have tried to find a reliable identity, a viable answer to the question "who are we". If there was a simple answer to that question it would have been found long ago. After all we have been around since Lucy walked on the planet and have had plenty of time to get to know ourselves.

For thousands of years we were the Children of gods or God. In modern times this conviction has lost ground to another: We are now the Children of Nature.

A substantial part of this modern theory is the notion that we are the eyes and ears of nature. It is through us that nature has become aware of itself. I came across this idea already in high school. We had a history teacher who loved to make digressions from the history course. One of the students dared to ask him: What is the meaning of life?

The teacher looked very pleased about being trusted with such an important question, leaned back in his chair, put his hands behind his head and looked up at the ceiling.

I remember that his face reddened as if the question went straight to his heart.

"In my opinion, there is only one answer", he said after a moment of silence. "Through humankind the universe has become conscious of itself, can understand what it is and maybe also, in years to come, why it is here. That is the meaning of our lives. Knowledge."

We nodded thoughtfully in our desks. The explanation seemed plausible.

The same idea crops up in the writings of many scientists and philosophers, modern as well as ancient. It is frequent in the books mentioned in this text. Many marvel at the transition of the "cosmic stuff" into consciousness and the pursuit of knowledge.

In Chapter 8 I quoted a physicist dwelling on the idea of a cosmos that has "stirred to life and through us human beings starts to wonder why it exists at all". (UD).

For the past 150 years it has been common to view consciousness in this manner. Many of the followers of Charles Darwin expressed themselves in similar terms. Julian Huxley, biologist and an elder brother of Aldous Huxley (*Brave New World*), regarded consciousness as the result of an evolution several million years old.

> As a result of a thousand million years of evolution, the universe is becoming conscious of itself, able to understand something of its past history and its possible future. This cosmic self-awareness is being realized in one tiny fragment of the universe – in a few of us human beings. Perhaps it has been realized elsewhere too, through the evolution of conscious living creatures on the planets of other stars. But on this our planet, it has never happened before. (Transhumanism 1957, Wikiquote)

The same thought can be found on the pages of *My Life*, the autobiography of C G Jung, where he invokes "the service that man can do to God, that light may arise from darkness to make the Creator conscious of his work and the human being of itself."

You might think that all these persons, and there are many more, who deem such an awareness crucial, should examine the phenomenon with the utmost care. They do not. They simply put forth the idea as if no further explanation or scrutiny is needed. Cosmic self-awareness is to them something self-evident. It is not a problem and needs no discussion.

But all those who talk of self-awareness should ask themselves: What happens when a self, any self, becomes aware of itself? In what way will this self be present to itself?

We really have to take these questions seriously and ask ourselves: can a self be aware of itself such as it is? Can I see myself the way I am? Is an exact image possible?

Once more I have to bring up Jean Paul Sartre who seems to deny such a possibility. Consciousness does not include being, he says in *Being and Nothingness*. (I mentioned this in Chapter 9, but will quote the passage once more to remind the reader).

> Consciousness is, because it is producing itself, which is due to the fact that its being is consciousness of being. This means however that the doing (le faire) maintains the being; consciousness must be its own being, it is never maintained by being; consciousness maintains being in the heart of subjectivity, which consequently means that it is inhabited by being but it is not being; *consciousness is not what it is.*

"It is inhabited by being but it is not being" seems to me to be the heart of the problem, together with "consciousness is not what it is", a phrase that catches the illusive character of the phenomenon. At the same time I think there is more to consciousness than this.

Again: what happens when a self becomes aware of itself? Sartre says that "consciousness is not being". But self is being. So consciousness does not include being?

A self must have certain characteristics, certain qualities. It must at least be *something.* And *this* is the self – and *no other* – that is subjected to self-awareness.

This is our starting-point. Now, if a picture of a person is exactly the same as this person, will he or she be able to see the picture? If the map of a forest is exactly the same as this particular forest, will we see the map? Of course we do not. Does not this also apply to human self-consciousness? Must there not be a deviation from the self in order for awareness to exist? I think so: what we are conscious of can never be the same as what we are, even if we always and irrevocably are what we are. In other words: *a self can only be aware of itself as it is by being aware of what it is not.* Consciousness creates a deviation, an invisible "not".

This might be called the law of self-consciousness. The deviation makes us conscious. But it is not any deviation but one *that diverges without doing so, a change that changes all without changing anything at all.* The self is still there, and must be, because otherwise no deviation can take place. The deviation makes the self necessarily invisible, never discovered in the discussions about consciousness, language and human nature. (It might be helpful to think of Wittgenstein's "what belongs to the essence of the world language cannot express).

The deviation is effortless and as natural as the act of breathing. When we tell ourselves that we are conscious of ourselves we are in fact conscious of something other than ourselves, *even though we always are ourselves.* The self is present to us through a deviation.

The only thing at our immediate disposal is the deviation. Consequently we call it "I" or "what we are", and rightly so. And we say such things as "I am who I am" or "this is me, you have to accept it". "This above all; to thine own self be true" is Polonius advice.

What causes the deviation? What changes the self without changing it? Language. It is the tool, the prerequisite of human consciousness. Without language we would not have a self at all, with it we have a provisional self that is not provisional. Language pushes aside *what is* by showing it as *what is not,* (while it still remains the same). In this respect the word is closer to us than anything else. It is constantly there to change everything without changing anything at all. Language therefore, could never separate us from ourselves or from what is. In itself it is the "essence". It is being. Consciousness, with the help of language, is at the same time what it is not and what it is. As far as I can see Sartre never reached that conclusion.

We are immersed in language in the way Wittgenstein has showed. It is impossible to look at language as a detached object only. Those who want to answer the question "what is language" have to do so *with* language: write, talk, lecture. It is like measuring a tape with that same tape. We are in the hands of language while it is in our hands.

This mechanism makes it impossible to doubt the language-games. It gives us complete, or almost complete, security. Therefore it is wrong to draw the conclusion that the deviation caused by human consciousness/language turns reality into some kind of illusion. The contrary is true. Language makes reality real albeit with an "invisible not".

What applies to our relationship with ourselves also applies to the world. We cannot be conscious of its essence, just like Wittgenstein says. On this faulty assumption science has built its empire during the last 400 years without being more than marginally disturbed. Our discoveries can only be concerned with the world *such as it is without being so.* Even so, these discoveries can never be called illusions. On the contrary: science creates reliable maps, and images, but it is no more than a map-maker. Science is a way of living, of relating to our surroundings, but it has nothing to do with the truth. It can only mimic, never give us the truth about ourselves or the universe. A map of Oslo can be better or worse (and really misleading if it is labelled a map of Stockholm), but it can never be the truth about Oslo if it is not identical with Oslo. And if it is, we cannot be aware of it. The scientific knowledge is based on models, but the prerequisite of the model is the deviation. Without it no models.

From this perspective Richard Dawkins is obviously wrong (*The Magic of Reality*).

> Science thrives on its inability – so far – to explain everything, and uses that as the spur to go on asking questions, creating possible models and testing them, so that we make our way, inch by inch, closer to the truth. If something were to happen that went against our current understanding of reality, scientists would see that as a challenge to our present model, requiring us to abandon or at least change it. It is through such adjustments and subsequent testing that *we approach closer and closer to what is true.* (My italics)

Models do not close in on truth simply because they remain models. Consciousness puts a stop to the ambitions of Dawkins and other scientists to reach their cherished goal. Science, in all its limitations, can still be helpful, even if it should not be mixed up with truth.

The difference between human and animal consciousness is huge. The authors of *What is Life* (Karolinska Institutet) maintain that human consciousness simply is an extension of the animal one. "At present research indicates that man has more of it (consciousness), but like it is with language and the use of tools we cannot assert that it is uniquely human." This is not so. Animal and human consciousness have very little in common even if our brains are similar. Animals are not conscious *of the world.* It does not exist to them as it does to us, because they *are* the world, connected to it with a direct link through their instincts, while we are connected indirectly by our language instinct in the way Wittgenstein describes it.

Primatologists have found that there is no "why" in the world of chimps and other apes. If a change is made in an experiment and routines are upset, they feel no urge to find out why. They cannot do this because they are nature: they are what they are. You can only ask "why" if you are disconnected from nature without being so. If you are nature you do not have to explain anything, or ask why. But that is not our way. The "why" is exclusively human.

Animals are the answer to a question posed by the environment. We are the answer to our own question. Consequently animals are not inclined to ask questions like “who are we” or “what is life”, because they answer such questions by living the answer.

It is difficult to understand how ”what is” can be something else without ceasing to be “what is”. The problem is one of looking both ways: understanding the presentation that language makes of “what is”. This is the same as realizing that consciousness does something with the world and with us without doing anything at all. The authors of the Paradise Myth in Genesis discovered the problem but did not solve it, seeing only one part of it: the (lingual) expulsion from Paradise. Therefore the whole story resembles a fairy tale. A true story of expulsion would show how we have been cast out while still remaining in the same place: Consciousness/language dropped an atom bomb on Paradise without stirring a leaf.

To be conscious of myself I have to be an object to myself even though I am a subject, which resembles a transformation. It rather seems to exclude consciousness. And if there is no difference between the object and the subject there is nothing to be conscious of. *Those who are conscious must be different from themselves, while they still are themselves.* In this process language is the sole instigator. We get a word for ourselves. Consciousness opens the world by closing it, blinds us with our own eye-sight. We are expelled and still belong.

If we understand our consciousness (and language) in this way we also get an explanation to the evasiveness of the “I”. Many philosophers and scientists argue as Daniel Dennett that the brain, due to the way it works, excludes the possibility of a central “self”. In that case the “I”, or self, would be a kind of illusion, a benevolent mistake. To presume the existence of a self would take us back to Descartes and his dualism, where a small being, a self, watches an inner screen, a scenario Dennett calls “the Cartesian theatre”. Susan Blackmore nicely outlines this idea in *The Meme Machine.*

> The theatre metaphor may do more harm than good to our thinking about self and consciousness. Dennett (1991) argues that although most theorists now reject Cartesian dualism, they still secretly believe in what he calls the ‘Cartesian Theatre’. They still imagine that somewhere inside our heads is a place where ‘it all comes together’; where consciousness happens and we see our

> mental images projected on a mental screen; where we make our decisions and initiate actions; where we agonize about life, love, and meaning. The Cartesian Theatre does not exist. When sensory information comes into the brain it does not go to an inner screen where a little self is watching it. If it did, the little self would have to have little eyes and another inner screen, and so on.

If the idea of a self *that is not what it is when it is what it is* is taken into consideration, dualism has to go. We are always what we are – no dualism here – even though consciousness and language cannot express the truth about what we are. The "I" is there as if not, but still impossible to eliminate as some philosophers try to do. It is not a mistake, not even a benevolent one, because it is made possible by what is, by what disappeared without disappearing at the very moment of self-awareness. It resembles a map, a model necessary for existence.

If this is right we must take farewell of many previous theories about the self, give up our hopes of finding a self in the brain. It can only be there as if it was not there. You might say: The "I" is the model of the self, being and not being the self at the same time.

To summarize the discussion about the self: the "ego-theorists" who believe in a continuum called self ought to be as wrong as their opponents who believe in its non-existence. David Hume (1711-1776) was among the latter. According to him, we just have access to our experiences, the experiencing self is nowhere to be seen. The only religion that does not acknowledge a central self is Buddhism. The idea of a self is false, Buddha reckoned, misleads us and causes suffering. Kierkegaard on the other hand points to an eternal self behind our attempts to cling to a superficial one. He talks of the possibility of being ourselves in "our eternal validity". But all these ideas are incompatible with the model of the self *that is not what it is when it is what it is.* The self is not accessible to consciousness, not by meditation or in any other way, but that does not mean it is non-existent.

Adam Zagajevski, the Polish poet, depicts the versatility and the evasiveness of the self in a long poem, of which this is the concluding part.

> ...it is always I, every morning opens a brilliant new chapter, that it is too feeble to end, this is I in the street, at the station, I listening to the child weeping, the laughter of the students, the

> whistle of the starling, I the ignorant, I the uncertain, I the thirsting, the waiting and wildly elated, I who understand nothing, who answer the proposals, doubts, try to start all over, take refuge in conversation, in despair . . . I at the museum, by the sea, in the market place of Krakow, longing for the moment that will not come, hiding like the mountain crests a cloudy afternoon, at last insight strikes, and suddenly I know everything, know that this is not I.

With the two words that end the poem Zagajevski opens for an endless search for identity, in which we now and then will be struck by the same insight: this is not I.

Now we are also able to understand why there is subjectivity, even though it is not found anywhere else in nature. We can answer the question how humanity is possible in a non-human universe. Language/consciousness did it, as easily as switching on a light.

"The enigma is the essence itself and our own subjective consciousness", Ulf Danielsson writes in *The Best of Worlds.* This enigma dissolves as did dualism if *we are not what we are when we are what we are.* Then and only then will it be possible to explain why human beings who all resemble each other, whose bodies are constructed in the same way, who have identical brains, live with opposing ideas about themselves and the world, as many as the stars, varied as the life-forms evolution has created. Now we can explain why "an impersonal, unreflective, robotic, mindless little scrap of molecular machinery" (Daniel Dennett) can cause superstition, animism, religion, philosophy, science, music, empathy – human subjectivity. If we had been directly linked to nature subjectivity would never have existed.

Germund Hesslow, neuroscientist at Lund University, is amazed at the subjective capacity of building alternate worlds within ourselves.

> It is a remarkable fact that we can experience the external world that we can see, hear, feel and smell things around us. Even more remarkable is that similar experiences can take place within ourselves, that we can talk about things solely in our imagination, that we can listen to a silent conversation in our heads, that we can have thoughts not only about the world around us but also about our own thinking. This 'internal world' of thoughts and feelings can

> be almost as real as the external world, but it is so radically different that many people have asked if it is at all within the reach of science . . . No matter what the answer is we must concede that the inner world gives rise to many intellectual problems. (*The Brain*)

Indeed! By noticing the deviation that consciousness creates we can understand the origin of this 'internal world'. To beings who "*are not what they are, when they are what they are*", "what is" disembarks from itself and becomes something totally different without changing a bit. The divergence inserts an invisible "not", or "un", into all possible propositions about ourselves and the world, making them provisional. But the truth is not lost in this process. It is only removed from our consciousness and remains as it always will remain, closer to us than anything else, as language, in the creation of ourselves.

We are the word and can only understand ourselves from that point of view. If we are misled there is always the possibility of recovery. Language gleams with immortality.

> My house consists of matter, and physics can answer almost all questions about its qualities. The house in my dreams, on the other hand, does not consist of matter. It is large but occupies no space. It is beautiful but reflects no light. Accordingly, it has many qualities that are physically impossible. But how can anything that is physically impossible emerge from a lump of matter, my brain, which is ruled by the laws of physics? What does the dream house actually consist of? This is the philosophical body-soul-problem. (Germund Hesslow in *The Brain*)

Now we have solved that problem. The dream house is made possible by the subjectivity that is created by language/consciousness. It is there because "what is" appears as *what it is not.* A simple language process creates the soul, the "I" and human reason!

The idea of a "soul" is a mirage pointing in the right direction. Language in the form of being ourselves, understood as our own bodies, is the soul. It is always there.

If we look at ourselves in this way we can understand why human identity is provisional, prone to incessant change. A horse cannot believe that it is a cat, or a fish that it is a bird. But human beings can believe that they are worthless or that they are the gifts of God.

There seem to be no boundaries to human inconstancy. This would not be possible if we had a fixed nature and certain recurring qualities. If we cannot get hold of "what is" we have to construe ourselves. And those who construe themselves leave openings to doubt, to feelings of anxiety, insecurity, loneliness, imperfection, hesitation and weakness.

There are many options available to those who construe themselves, which gives rise to an exclusively human anxiety, one that is not found anywhere in the animal world.

"Anxiety can be compared to vertigo", Sören Kierkegaard says. "The eye that stares into a gaping abyss will give rise to vertigo". The abyss can be understood as the illusory absence of a self, the necessity to construe oneself in this absence.

If anxiety is caused by language and consciousness showing the self as *what it is not when it is what it is* it should be possible to free ourselves from anxiety by making "what is not" vanish into "what is". Then we would always be what we are. But I have just shown that this is impossible. We would not be conscious of ourselves if that happened.

In the last chapter I will try to show how "what is not" can still become "what is", how theory and practice, the word and the body might become one. After all, we are the word. We left paradise without ever leaving it. That is the human tragedy.

Chapter 12
The invisible 'not'

If I am right about the duality of self, which is no duality at all, notions about our identity must be provisional and temporary no matter how reliable they may seem.

You might object that if provisional ideas are the only ones available, we should regard them as stable and abandon all hope for something else. Why presuppose the existence of "what is" or a "self" when we cannot say, "this is the self"? Well, what happens if we do not presuppose? Then we are stuck with the misconception that it is possible to say, "this is we, this is the truth, this is human nature", phrases that can be but provisional.

If there is such a duality, it will be easy to understand why human identity has shifted radically over time and still does. Some people say that we are evil, some that we are basically good, some that we are mixtures of good and evil. If we take the fleeting character of the self into account, we will understand why Rousseau believed that man is good in his natural state, why Lasse Berg thinks that man was peaceful and social as a hunter and gatherer and why Rolf Edberg tells us that it was a "murder ape" that climbed down from the trees and populated the earth. They pin truth on the representation, the provisional form.

All attempts to finish the sentence "humankind is . . ." in other than a provisional and temporary way are doomed to fail. It does not matter who construes the sentence, were it God himself. When Richard Dawkins says "we are survival machines for our genes" there is no more truth in that than in an everyday drunken comment at a bar counter. The introductory phrase "We are . . ." "Humankind is" . . . can never be followed by permanent truth.

There are abundant examples in literature of "true identities" that in fact are provisional. Some are provided by what is called "evolutionary psychology" and its talk about "universal human psychological architecture", a field of research especially popular in the US.

Psychologists like Steven Pinker, Leda Cosmides and John Tooby all believe that they can explain human behaviour by linking it to the evolution of our species. In an essay on the net (Evolutionary Psychology – a Primer) Cosmides and Tooby depict our common origin.

> ...all normal human minds reliably develop a standard collection of reasoning and regulatory circuits that are functionally specialized and, frequently, domain-specific. These circuits organize the way we interpret our experiences, inject certain recurrent concepts and motivations into our mental life, and provide universal frames of meaning that allow us to understand the actions and intentions of others. Beneath the level of surface variability, all humans share certain views and assumptions about the nature of the world and human action by virtue of these human universal reasoning circuits.

They tell us, in sophisticated scientific jargon, that all human beings are standardized as far as thinking and experience go because our "reasoning circuits" are the same.

They even surmise that "beneath the level of surface variability, all humans share certain views and assumptions about the nature of the world". This must be astounding news to many scientists and philosophers at least to all those who fight over the essence of the world and human identity. Why do they disagree when we all stem from the same evolutionary mould and have coinciding assumptions? How can diversity be explained through similarity?

The books of Steven Pinker are imbued with the same idea. In *The Language Instinct* he contends that beneath the surface of all languages there is a common language. To prove this he also tells us that even if he does not understand all languages he still knows what people using a strange language are talking about. Just as the mental grammar is universal, there is a universal design to the mind, which gives us hope for a future of brotherhood.

> Just as there is a universal design to the computations of grammar, there is a universal design to the rest of the human mind – an assumption that is not just a hopeful wish for human unity and brotherhood, but an actual discovery about the human species that is well motivated by evolutionary biology and genetics.

Thanks to evolutionary biology we have reached a new and deep understanding of our human roots! The problem is that the inert general public has not yet caught up with these truths, dragging their feet in fights over what is clarified by science. Not even highly trained and educated politicians know that they in fact are brothers, not adversaries!

The Language Instinct ends on a note that elevates Pinker to a god-like position, sitting on a cloud overlooking humankind, knowing all its ways and hidden thoughts.

> Knowing about the ubiquity of complex language across individuals and cultures and the single mental design underlying them all, no speech seems foreign to me, even when I cannot understand a word. The banter among New Guinean highlanders in the film of their first contact with the rest of the world, the motions of a sign language interpreter, the prattle of little girls in a Tokyo playground – I imagine seeing through the rhythms to the structures underneath, and sense that we all have the same minds.

By a deft or even cunning use of what is not scientifically proven, Pinker convinces himself that he knows what he only imagines to know. I am sure that Pinker would not understand a person from a society of equality even if they spoke the same language.

We might have the same evolutionary background and similar wiring in our brains, but there is still not a single view or notion that I have in common with Cosmides, Tooby or Pinker. What they propose is in my view a trendy, provisional map of what it means to be human. It will change, because *we are not what we are, when we are what we are.*

The same critique can be applied to a similar idea: that the hunting and gathering stage of our history "designed" us. This is a cornerstone of evolutionary psychology. Cosmides and Tooby think that the circuits of the brain are "designed to generate behaviour that is appropriate to our environmental circumstances". It must be somewhat of a mystery to them that these circuits are more than helpful in destroying these environmental circumstances.

They point to brains that "generate sophisticated behaviour that is sensitively tuned to its environment". I am not aware that such brains exist. Where are they?

> Generation after generation, for 10 million years, natural selection slowly sculpted the human brain, favouring circuitry that was good at solving the day-to-day problems of our hunter-gatherer ancestors – problems like finding mates, hunting animals, gathering plant foods, negotiating with friends, defending ourselves against aggression, raising children, choosing a good habitat, and so on. Those whose circuits were better designed for solving these problems left more children, and we are descended from them . . . Our modern skulls house a stone-age mind. The key to understanding how the modern mind works is to realize that its circuits were not designed to solve the day-to-day problems of a modern American – they were designed to solve the day-to-day problems of our hunter-gatherer ancestors. These stone-age priorities produced a brain far better at solving some problems than others...

Here you would expect a more detailed catalogue of these day-to-day problems of our ancestors. But all Cosmides and Tooby manage to present is a theory about snakes: "that it is easier for us to learn to fear snakes than electric sockets, even though electric sockets pose a larger threat than snakes do in most American communities". To me this is rather amusing. My grandchildren are all aware of the danger of electric sockets (the Swedish standard is 220 volts). To them snakes are likable living creatures among scores of others.

According to Cosmides and Tooby, our minds can be compared to Swiss army knives, equipped to meet a wide range of problems. As Daniel Dennett puts it: "We have a collection of gadgets, all pretty good and readily adaptable for new purposes today". But "every now and then, we discover curious gaps in our competence, strange lapses . . . that explain the machinery that underlies the glittering facade of culture". Dennett, Cosmides and Tooby would be more convincing if they showed us exactly what those "lapses" are. But such a list does not exist, because it would topple the Swiss army knife metaphor.

My "duality-model" of human consciousness and language also elucidates the whole field of primatology and all the attempts to link human beings to the larger apes.

The Swedish anthropologist Lasse Berg visited the primatologist Frans de Waal at Emory University in Atlanta, USA, to learn about this kinship.

Dawn over Kalahari contains an account of his talks with de Waal and of his encounters with the apes.

de Waal points to the behaviour that links us to chimps such as laughter, facial expressions and other non-verbal communication. "Many parts of what is at the bottom of humankind is also at the bottom of the chimp's chimpansity", is Berg's conclusion. He is even convinced that "it is difficult to point to anything in our nature that is particularly human".

Berg sympathizes with de Waal and contributes some reflections of his own. There is only one thing that makes him doubt the proven relationship between humankind and apes. Maybe human consciousness is unique after all? It does not seem to be ape-like. His premonitions point in the direction of language. Maybe "the distinctive character of human beings has a lot to do with language"? To know more about the rise and function of language he pays another visit, this time to Terence Deacon, American neuroscientist and anthropologist.

Deacon tells him that the special life of our ancestors "demanded extraordinary tools". Symbols are such tools and so is language. But this is the very misconception of language – that it consists of symbols and is a tool – and thus scientists are guided by a misconception in their work with the apes. Scientists do not take the language predetermination into account. They do not see that if animals were to have a language like ours, they would, taken predetermination into account, be able to create themselves and the world around them.

The language of animals, in so far as they have one, is what it is, but our language is not what it is, when it is what it is. (Consciousness is not what it is). What the scientists call language is nothing but a signal system that apes are capable of using to a certain extent in connection with food. Susan Blackmore has studied many of these experiments and she concludes that attempts to teach language to other species are close to complete failures. "Whereas children just seem to absorb the words they hear and turn them into language, chimps have to be coerced, and rewarded to learn just a few paltry signs."

The misunderstanding of language is evident, when a scientist like Sue Savage-Rumbaugh makes a statement about the result of her experiments. "You cannot avoid the conclusion that Kanzi (a bonobo) has a fundamental language." The very words give her away: you cannot *have* our type of language without being it. Those who believe that apes can have or get a language must also believe that they have a culture. Or shifting identities. The one follows the other. Many authors in this area think of apes as cultural beings. "Chimps are cultural beings who discover, develop and teach new techniques and behaviour". (Jonathan Lindström in *The Mystery of Consciousness*). Do you have a culture, if you use techniques like fishing termites with a straw? Or if these techniques change between flocks? It seems to me that chimp culture is as different from human culture as are our languages. Apes have had millions of years of "cultural" development without getting further than using sticks to pick bananas.

The misconception of language and culture leads to clumsy mistakes like this. But they are easy to remedy. Every language institute should have a Wittgenstein department to avoid these embarrassing misunderstandings. The Austrian philosopher has once and for all enlightened humankind on the issue of language and its function, but humankind does not seem to listen. I always look at the reference list of a book to see if Wittgenstein's name is there. Very seldom it is, especially in American books. Is he overlooked in the US?

If you try to say "humankind is . . . " you are bound to insert an invisible "not" in the sentence. "Humankind is social" (not), "Humans are solitaires (not) ", "Humankind is good" (not), "Humankind is evil"(not), "Humankind is both good and evil" (not) and so on. You cannot even be sure of what seems to be an axiom of human behaviour: "Perfectibility remains, as ever, an illusion." (Ian Tattersall in Becoming Human).

There is no stable identity for humankind (while there is one) and that is also the reason why all attempts to find one fail. Berg and other anthropologists can cross the Kalahari-desert innumerable times to prove our identity – they will never succeed. No one will.

According to Steven Pinker male psychopaths in the US are around 4 per cent (Other sources say 1 per cent worldwide). In US prisons around 25 per cent are psychopaths.

It is a mystery to Pinker and other evolutionary psychologists why the percentage is so high. According to evolutionary theory, it should be the other way around: the tight social life of hunters and gatherers made it difficult for psychopaths to spread their genes. Propagation was on the other hand facilitated for those who cooperated and provided for others. So, if "our modern skulls house a stone-age mind" the high percentage of psychopaths *should* baffle us. There ought to be very few psychopaths or none at all! You may also wonder why the percentage is different in Sweden and the US. Do we not stem from the same evolution?

Berg cannot find a solution to the problem. "It is an evolutionary mystery how this change of genes (towards psychopathy), if that is what it is, has been able to catch on." And he goes on banging his head on the evolutionary wall: "The capacity for goodness, selected during millions of years of equal group existence, may not have the same advantage in the hierarchical societies we recently invented." The very words point to another problem: how hierarchical societies could be invented given our social, genetic heritage as hunters!

Biologists and anthropologists are very often blind to the invisible "not", which would threaten their sciences. And they share this blindness with large parts of the cultural establishment. We have probably all heard somebody say, "literature gives us honest and deep insights into what it means to be human". These particular words are from a debate in a daily newspaper. But "what it means to be human" cannot be described by literature. We can only write books – and these must be written – about what it means to be human in our type of society, in our time and in historical times that we know of. The truth about what we really are is inaccessible to us – but it is not the same as saying that it does not exist.

I have already, from another point of view, dealt with the misconception that "humankind is a product of evolution". The invisible "not" is there too, as it is in expressions such as "humans are social beings", "humans are biological creatures", "humans have dark sides", "humans are imperfect", "humans are perfect" and so on. All these expressions are not to trust. Some words from *Being and Nothingness* by Sartre reflect this:

> The sincerity as a constant ambition to cling to oneself is, in its nature, a constant striving to betray the solidarity with oneself; one is liberated from oneself through the very act with which one makes an object of oneself. *Constantly to make an inventory of what one is means a constant denial of oneself.* (My italics)

All these statements about what a human being is add up to a denial of what we are. When I came to this conclusion – the same as Sartre's – I wanted to find out how widespread these inventory expressions really are.

They abound in the writings of Donald E Brown and others who believe in the theory of Universal People. But are they also frequent in fiction?

To find out I consulted twenty of our latest Nobel Prize winners in literature and read their speeches from the Nobel Prize ceremony, all available at the committee's home page.

Around half of these speeches contained statements of the sort I have just mentioned, blatant attempts to once and for all establish the true identity of humankind.

Günther Grass thinks that there is a "rat-part" in human nature. And Gao Xingjian believes that literature can give "a true picture of the life of human beings". Imre Kértesz talks about "human nature" and "human conditions", stretching his knowledge beyond its limits. Seamus Heany underlines that humankind has a "sympathetic nature" and the world the opposite. V S Naipaul quotes Proust, who believed that you can find your "innermost self" by "pushing aside the world and the I that socializes with the world". This is of course a mere illusion. Hertha Müller talks about "the dark and light sides of the soul", which gives us yet another provisional identity. Orhan Pamuk confesses his belief in human nature by stating his conviction "that all men everywhere resemble each other". Pamuk also proclaims "literature as the most valuable item that man has created to understand himself". "Life is random in itself", says Nadine Gordimer (in itself life is something that we cannot talk about). To Mario Vargas Llosa life is a "treadmill". All these ideas are provisional and temporary.

The short catalogue just presented is not to be regarded as literary critique. These Nobel Prize winners have all written wonderful, astounding books that we should and will read over and over again. The point of my short inventory is that they are wrong in trying to pin down human nature, and so are millions of their readers who believe them. Idealistic ideas like these belong to what I described in an earlier chapter as the "doxa", the battery of notions that are designed to bolster the status of the writers in the eyes of the general public.

There are also, of course, Nobel Prize winners who stay away from these misconceptions.

One of them is J M le Clezio, Nobel Prize winner in 2008. He is fully aware of the limitations of literature, realizing that it is not common heritage. "How is it possible that on the one hand act as if nothing was more important than literature, while he (the author) on the other hand sees how people around him starve . . . He who wants to write for the hungry will be aware that only the satisfied have the tranquillity to notice him." Authors and the reading public must understand that "literature is a luxury for the leading circles of society, that it thrives on thoughts and ideas that are strange to most people . . . " Such perspectives puncture the glorified idea of literature as the guiding star of all of humankind.

> If there is a virtue that is absolutely necessary it is this: that his pen never pay tribute to the powerful, not even in the form of a remote tickle. And still, even if the author complies with this virtue, he should not feel cleansed from suspicion. His revolt, his refusal and his curses stay on the side of the barrier where the language of the powerful rules.

The perspective of le Clezio relentlessly reveals the political agenda behind such seemingly innocent expressions as "the nature of man" and "such is life". They are the flagships of reactionary politics, illusions of control, knowledge and authority.

When literature crosses the border of what is possible to say, when it talks about the world "such as it is" or human beings "such as they are" it betrays itself, becoming a reactionary tool in the preservation of society and thus in the oppression of human beings.

Chapter 13

The underestimated evolution

When Lasse Berg starts out to search for the roots of humankind in *Dawn over Kalahari,* he visits a number of well-known research centres. The researchers all seem to sympathize with his conclusion that "man is an animal whose different capabilities have been created during millions of years of evolution". They agree with Berg that we got our final "design" during the long period of a 100 000 generations when we lived as gatherers and later as hunters and fishers on the windswept coasts of southern Africa. After that period the time span has been too short to initiate decisive genetic change. "That life created our bodies, our brain, its thinking and our innate moral feelings", Berg concludes and echoes Darwin: "There is more in our innate behaviour that links us to other animals than separates us."

This is the crucial point. If we were clearly separated from the animal world, the whole theory of evolution would be at stake. What if we are both separated and related at the same time? Separated from the animal world through language/consciousness and still related through language in the form of "instinctive behaviour"? Berg and other authors look for a solely physical connection. Is there a possibility that they are misled by their firm belief in evolutionary theory in the same way that Ptolemy was misled by his geocentricism?

Is it possible that evolution turned a sharp corner with the birth of humankind? To Darwin's followers this is unthinkable. But if it is true that consciousness separated us from ourselves without doing so, the whole idea of a continuum seems to falter.

Then we are not simply extensions of nature with a tangible human nature.

I think something is missing that will up-end the whole Darwinian idea. As I have already mentioned the evolutionary psychologists Cosmides and Tooby compare the human mind to a Swiss army knife, a large but limited set of gadgets developed during the hunter-gatherer era.

These gadgets or tools made it possible – and still do – to deal with a number of problems and situations. Daniel Dennett supports the two psychologists in *Darwin's Dangerous Idea* when he surmises that "such rationality as we human beings have is the product of the activities of a host of special-purpose gadgets designed by natural selection". The idea is that if other species have one or two blades in their "army knife" we have hundreds.

The consequence of this idea is obvious. If our minds were like Swiss army knives equipped with a great number of gadgets, we would still, in the face of certain problems or situations that did not exist during the hunter or gatherer period, be at a loss. Everyone who has used a Swiss knife has experienced its insufficiency. How often do we stand at a loss depending on the insufficiency of our minds? Does it happen at all? I think these moments are rare, and if they happen, we will try new solutions, constantly rearranging our ideas.

How is a manifold development of cultures possible if special-purpose gadgets are involved in mapping out the road of humankind? Would they not move us in a few limited directions, prescribed by the number of blades in the mind's army knife?

Or are our minds construed in a way that does not resemble such knives at all? I have tried to show that our consciousness is not what it is, containing both "what is" and "what is not" in the same instant. Ptolemy introduced different auxiliary devices into his astronomy, such as epicycles, to make it viable. It seems to me that some Darwinists do the same thing. Richard Dawkins certainly does it in his book *The God Delusion*, in outlining the history of humankind. Even though he agrees with the anthropologists about the importance of the hunter and gatherer life, he makes an addition that is close to a denial of the evolution.

> "As an unexpected bonus our minds seem powerful enough to handle a much more complicated world-model than the mediocre, utilitarian one that our ancestors needed in order to survive. Art and science are runaway manifestations of this bonus."

How are bonuses possible? In the same book Dawkins compares natural selection to a stingy bookkeeper that always detects extravagance and minimizes it to avoid unnecessary spending. If so, how is it possible for natural selection to create bonuses? Where do they come from? Has the stingy bookkeeper emptied a bottle on his way home and forgotten about his calling? Bonuses are not needed for the sake of adaptation and should not be there. This is really astonishing: beside the blades of the Swiss army knife there would, according to Dawkins, be others with no function at all, waiting to be used in a "more complicated world-model". The blind watchmaker is opening one eye to foresee the future?

Why do Darwinists believe in "bonuses"? They should not. But they seek refuge in the idea because, as I see it, they have not understood evolution. During millions of years it created and adapted species through natural selection, to let the fittest survive.

But is this the limit of what evolution can accomplish, grinding the wheel one way only? Could it not accomplish something even more astonishing and grandiose? I think it did. If I were evolution I would try to create a being that is not limited to a specific environment, but can survive almost anywhere. *That would be the ultimate triumph of evolution.*

I think it succeeded in doing so. It followed a road that not even Darwinists could foresee in the creation of a species that could live in ice and snow as well as in deserts and arid places, on islands in the sea or in the steaming jungles of the Amazonas, in large communities of millions as well as in isolated settlements of a few individuals and even in space, and it did so with a simple device, all too simple for clever scientists to understand: language, the visible side of human consciousness. It took a couple of words to alter the world (without altering it), to create a brand new species connected and disconnected at the same time.

By making a determined identity undetermined (without letting go of the determination), evolution gave us the capacity to be anything rather than something, peaceful or belligerent, good or evil, contented or displeased. It gave us the opportunity of incessant development by opening the confinement of a "nature", making us boundless Swiss army knives. You might say that evolution trusted us with its keys. We have the possibility to destroy or adorn the world. In this respect we are no longer products of evolution. We *are* the evolution.

Evolution created animals that can recreate parts of themselves such as the hydras, bacteria that resist radiation and animals that for their survival have to lay eggs in other animals. But these were lesser triumphs: we are superior to all, by a long shot, because *we are what we are not, when we are what we are.* That is versatility at its peak.

Cats, magpies and beavers are what they are without being able to do anything about it. They do what evolution "has told" them. We are completely different without being so. Human consciousness and language seem to alienate us. This separation is real and non-existent at the same time. The problem is that the unity, the connection with the rest of nature is not upheld by anything that is directly discernible and tangible, but through what caused the division in the first place. Language. That is the human dilemma, if there ever was one: language is both the lock and the key. Our identity is not lost because language *is* this identity. As a mere instrument it leads us away from ourselves, banishes us from paradise. But as a creator of identity, something that cannot be objectified, it keeps us there.

We are one with all living things, but not primarily through our dependency on air, water and soil or because we are physically related to other mammals. The relationship is established by language. Therefore you might say that animals are the answer to a question posed by the environment. But we are the answer to a question posed by ourselves.

If we only are able to understand ourselves in terms of "what we are not, when we are what we are" many scientific "truths" will have to be reconsidered. The road to a permanent "human nature" will be barred, as well as any attempt to say, "this is the truth". Science will have to realize in a more serious manner than before that it deals with models, symbolic knowledge that probes and examines the essence of the world without doing so.

To reach a higher level scientists have to shift from being parachutists into a dark world to becoming participants of a fully illuminated one. And that is not done without social change; a twofold movement towards equality and the development of a new language.

What has to go, and will go, are Standard Models developed during the 20th century. One of them is the Standard Social Science Model. (Dennett's *Darwins dangerous idea*).

> Whereas animals are rigidly controlled by their biology, human behaviour is determined by culture, an autonomous system of symbols and values. Free from biological restraints, cultures can vary from one another arbitrarily and without limit . . . Learning is a general purpose process, used in all domains of knowledge.

Dennett dislikes this model so much that he has to say, "this, of course, is wrong, wrong, wrong." He modifies it with his own 'Nonstandard Social Science Model'.

> Whereas animals are rigidly controlled by their biology, human behaviour is *largely* determined by culture, a *largely* autonomous system of symbols and values, growing from a biological base, but growing indefinitely away from it. Able to overpower or escape biological constraints in most regards, cultures can vary from one another enough so that important portions of the variance are thereby explained . . . Learning is *not* a general-purpose process, but human beings have so many special-purpose gadgets, and learn to harness them with such versatility, that learning *often* can be treated as if it were an entirely medium-neutral and content-neutral gift of non-stupidity.

Dennett is hovering between two opposing views without ever landing. Cultures are restrained by biology as well as free of it, he says. They can overpower biological constraints but only to a certain limit. Why not show us the limit and what the constraints look like! To do this *before* the conclusions are drawn would have been honourable science. But Dennett does no such thing. He gives us the result of the game without ever playing it.

"Overpower" also seems to be a very strange word in this context. How did primitive human beings overcome "raw" biology? What kind of willpower was required?

It seems to me that Dennett envisages a life of basic biology that never existed. In case you have forgotten: when humankind enters the stage, so does language, society and culture.

"This is the model I have argued for in this book", Dennett says. I can tell him that his efforts are in vain. The two models are wrong in exactly the same way. Their engineers presuppose that they can tell the truth about the origin of human behaviour. But that is not possible. There will never be a Social Science Model that is permanent and true. The truth about ourselves and the world (as I will show in the last chapter) can only be expressed by our being it, not only having it. Theory that does not overlap with practice is fallible.

Why is there such a high degree of conformity among scientists? Is it because reality is fundamentally the same everywhere (as well as their minds), or is it because they have gone to similar schools and live in similar societies? Because they are children of the same tradition? There is no doubt in my mind that the last alternative is the right one. Reality never dictates anything about the truth of existence, because our communication with it is solely through models. And the models are our responsibility. To really know something, and move away from provisional knowledge, we have to know the essence of the world and of ourselves. But language makes that impossible, as impossible as for a geologist to examine the ground under his feet without moving. Still, language is always with us, presenting the essence without doing so. If we take language seriously it will give us all the answers.

Other models will also become redundant, among them the idea of "evil".

In *A Small Book of Evil* the Swedish author Ann Heberlein studied several evil acts in order to explain them and find their causes. "My interest in evil is almost perverse", she admits, which motivated her to write the book. In spite of this her conclusions did not clarify the problem. In her book humankind never unites, but seems inevitably divided.

On one side there are those who fight evil, and on the other the perpetrators. The author herself belongs of course to the righteous side. But in order not to look too untarnished and good she accepts a portion of the guilt, telling us of her passivity when it comes to revealing and denouncing evil. Because she is passive, she is also guilty and related to the perpetrators. She laments: "We who are spectators. We who stand on the side. We who choose not to listen. We who look the other way. Because it is simple. Comfortable."

Her book about evil, like many others, omits the real problem: that of discerning the roots of evil. We can all identify atrocious deeds. But where do they come from? Where is evil in the well-fed and orderly suburb? Or at the smoothly run office of a large company? The difficulty about evil lies in defining what it really is and where it comes from.

Ann Heberlein is quite often seen on Swedish Television. In one program, acting as a specialist on atrocities, she stated that "there is a seed of evil in all of us". The other participants, the same kind of amateur philosophers, nodded in agreement. But if my model is correct there are no such discernible "seeds". What we are is not accessible to language or consciousness, not through meditation, not in any way, because of the construction of consciousness. We are *not* what we are, when we are what we are. There is consequently no part of us inhabited by evil. The idea is unfounded, a grandiose misunderstanding.

Heberlein goes on to say that it is difficult to explain what evil really is and why people act against their better judgement. It is certainly difficult if you use a static model of a human being. Evil acts are on the other hand possible to explain using my model. They are there because we are two-dimensional without being so, severed from ourselves while still being ourselves. We are alloys of “what is” and “what is not” and this is the root of evil.

So who is evil, who is not? According to my model: everyone and no one. Baruch Spinoza comments on the problem in a way that I wholly agree with. In one of his letters he says that “evil has no positive existence”. “Because it expresses imperfection, sin or evil cannot consist of anything that expresses reality”, Spinoza writes. In other words: evil exists as if it did not exist at all, filling up the part of “what is not” to the brim. This approach to the theodicy problem solves it completely. Evil is a denial of “what is” embracing “what is not”.

The existence of evil must be sought in a different place than human nature, *in the foundation of ourselves by means of language.* Do not forget that language is closely knit to society, carrying its structure within itself. Sin or evil gets its momentum from the construction of society. And language. That is why the roots of evil are difficult to locate.

If we are not what we are, when we are what we are, we can minimize evil by becoming more of what we are and less of what we are not. Diminished alienation and increased self-knowledge – that is the key. How this is done, I will presently show.

Chapter 14

A true model

Certain notions abound in the works of authors with a scientific background. The common idea is that science builds reliable models of nature making predictions possible.

This is the opinion of Richard Dawkins in *The God Delusion*, where he writes that he regards the "endeavours of humankind to understand the universe as an enterprise of model-building". (Apart from being a standard answer it is an interesting approach to let scientists represent humankind. The claim is part of the "doxa" described by Pierre Bourdieu).

"We build models of the world and compare them to reality", says Ulf Danielsson.

The same approach has been used incessantly during the past 400 years, starting with Rene Descartes and Francis Bacon. Both of them, as well as innumerable followers, thought that reality is accessible "as it is". But as I have shown this is both right and wrong.

Bacon, called the father of science, presented his epoch-making theories in *The New Organon*, the new method. During centuries of superstition subjective notions had blurred the vision of humankind, he argued, covering our eyes like a spider's web. If we followed his new method our eyes would look untainted at the world. Nature would present itself to us "as it really is". It was in fact the highest possible achievement of humankind to embark on this new mission: "the very contemplation of things as they are, without superstition or imposture, error or confusion, is in itself more worthy than all the fruit of inventions."

Religion was crucial in Bacon's attempt to launch a new model of the world (even though he regarded Catholicism as a form of Christian superstition). Descartes and Bacon realized that they would never be successful without referring to God. And the crucial argument was obvious. After all, God had created nature as well as the human senses. That guaranteed the understanding of nature. How could god-created minds *misunderstand*? We should read nature's book without any fears of error because God himself is the author.

The investigation of nature could be regarded as a form of worship, God and humankind talking to each other. Consequently, those who banned the new method were in danger of banning God. By linking religion and science in this way Bacon protected his theories and himself from clerical attacks. Even if a new understanding of the world emerged, obliterating the old one, it would be regarded as sanctioned by God. The new method, Bacon argued, was "the best remedy for superstition and the best possible nourishment for faith".

It seems as if knowledge was Bacon's ultimate goal. But it was merely a means in the exploitation of nature. "The true and lawful goal of the sciences is none other than this: that human life be endowed with new discoveries and powers." Bacon never tires of reminding his readers that there is no power without knowledge. Knowledge of nature's laws makes us capable of unlocking its secrets. "The roads to human power and to human knowledge lie close together and are nearly the same." "The empire of man over things depends wholly on the arts and sciences. For we cannot command nature except by obeying her."

We cannot go wrong, according to Bacon, because God will guide us. "Only let the human race recover that right over nature which belongs to it by divine bequest, and let power be given it; the exercise thereof will be governed by sound reason and true religion."

The same faith in sound reason and true religion permeates the writings of Descartes. In Discours de la Méthode (1637) he points out that the new method will make it "possible to gain knowledge which is very useful in life". He hopes that this new knowledge will replace the speculative philosophy that dominates contemporary thinking and show the way to a "practical philosophy which would let us use the natural forces . . . for all they are suitable and in that way make us masters of nature". Knowledge is power over nature.

The two philosophers were convinced that they were leading humankind into an era of enlightenment and bliss, once the web of superstition was gone. And Bacon certainly had a point in attacking the prejudices of his time. He found four classes of idols which "beset men's minds" preventing us from a clear view of nature. The first class he called *Idols of the Tribe,* which is the tendency of human nature to overestimate itself, blending its own nature with that of nature itself. Bacon says "the human understanding is like a false mirror, which, receiving rays irregularly, distorts and discolours the nature of things by mingling its own nature with it". The quest for objectivity cannot be described in a better way.

The second class is *Idols of the Cave*, pertaining to individual man, the private den or cave that "refracts and discolours the light of nature". It is caused by education, books, authority. This makes the spirit of man seem like "a thing variable and full of perturbation and governed as it were by chance". *The Idols of the Market Place* come from the "intercourse and association of men with each other". It thrives on the "ill and unfit choice of words". This abuse leads to "numberless empty controversies and idle fancies". Philosophical dogmas belong to the *Idols of the Theatre*. These idols are like "stage plays" building unreal worlds. The worst are those that mix philosophy and religion in an unwholesome way.

If people got rid of all these Idols, they would be able to understand nature, Bacon thought. His idea is comparable to von Wright's belief in "untainted language games". von Wright thought that we could "by weeding uncover language areas in their purity", expressing belief in the idea that the world could be seen "as it is" when cleansed from all prejudice and the ill use of words. Unfortunately, Wittgenstein showed that language in a way precedes itself, that prejudice or predetermination is necessary in order for us to understand anything at all. This necessary subjectivity is not observed by either Bacon or von Wright.

The subjectivity that Bacon wants to remove I would call secondary or unnecessary. But there is also a necessary subjectivity in the way we structure our consciousness through language. When Bacon turns a blind eye to this subjectivity, he obstructs understanding.

Bacon was too much involved with the assault on superstition to see this. He is convinced of his own clear-sightedness and is proud to introduce the new scientific method.

> For I am building in the human understanding a true model of the world, such as it is in fact, not such as a man's own reason would have it to be; a thing which cannot be done without a very diligent dissection and anatomy of the world.

This is the idea of modern science summed up in a sentence. The idea is contrary to the medieval view that interpretation and understanding should take precedence. Objectivity is the new, more viable strategy. What other way is there to fight superstition? Still Bacon does not see that his own recommendation is prejudiced too, involving a preselection of the value of objectivity. Being unaware of this, his method will churn out unwanted results, far from the original intention. The scientific view of the world will drag behind it a gaping emptiness, a misunderstanding of our place in nature that makes our own existence superfluous. Objectivity, Kierkegaard says, kills subjectivity, and if it does not, cheats in the operation.

Thus an unwanted result of pure objectivity is the strengthening of superstition in the form of religion. Bacon might have thought that he was winning the battle and that superstition would evaporate from the earth, but the contrary seems to be true. Religion is, in many parts of the world, stronger than ever. It is even widespread in countries that devote a major part of their education system to science! As I already pointed out around 40 per cent of Americans think that God created humankind 10 000 years ago. You might think that 400 years of Enlightenment would be enough to erase such ideas. But not even another 400 years would suffice. And that is of course because Enlightenment and science cannot explain existence.

Another consequence of Bacon's construction work is a self-glorification out of proportion. He compares himself to Prometheus, bringing humankind light.

In a biographical note he describes himself as extremely talented, a young man endowed with all the gifts that are needed to cast a light on nature so that all "what is most hidden and secret in the universe" will be revealed and clearly observed. He can manage this because his character displays the "will to seek, the patience to doubt, the disposition to meditate, the prudence to certify, the readiness to change of mind, the carefulness to decide and put into order". This grandiose exhibition of suitable qualities ends on the same oversized note: "I regarded my mind to have a kind of similarity, a relationship with truth."

If you regard yourself as an advocate of truth, it is easy to make mistakes, to overlook details and cut corners, guided by an inflated ego and an overestimation of self.

Bacon clearly fell into this trap. The oversized self-esteem might also have been heightened by his social situation as England's Lord Chancellor. In this office he literally resided over life and death, as death sentences required his signature. He is said to have had more than 70 personal servants, which would also emphasize his importance.

Did Enlightenment, pioneered by Bacon and Descartes, lead humankind into a state of bliss, peace, knowledge and security? I think the results are becoming more and more obvious. Knowledge that is power seems to tear "our house in cosmos" apart.

The two Enlightenment philosophers were not the first to embark on this deplorable journey; they were only part of a tradition with roots dating back to ancient Greece.

Aristotle regarded knowledge in a similar fashion to Bacon. Our senses are reliable, he thought, and therefore give us a correct rendering of the world and ourselves. von Wright, in his book *Science and Reason*, calls this "a belief in the intelligibility of the world order".

> The psychological foundation of the Greek philosophy and science can be described as a silent rather than outspoken belief that human thought by its own power, without the support of supernatural authorities, can understand the logos of things, that is the meaning and intrinsic order. You can call this a belief in the intelligibility of the world. It is reborn with the pioneers of the renaissance in the form of a belief that the "book of nature" can be read by humankind as long as we learn to understand its mathematical language. Such a faith or rather assurance that the existence is intelligible is the rational foundation of all that can be called science. This goes for our own era as well as that of antiquity.

The belief in intelligibility has accompanied humankind during thousands of years. From where does it stem? I am convinced that our instinctive use of language has a lot to do with it. Language, as I have tried to point out, precedes itself, thus rendering a deep sense of security. We do not even have to say that "this is a hand" as Wittgenstein points out, because the fact is already established. Language makes the world coherent. The idea that this reality could be explained seems perfectly natural: an explanation seems already in hand.

Another person with interesting views on intelligibility is Pierre Simon de Laplace, one of the pioneers of mathematical physics and the first one to predict the possibility of black holes in space. He is one of the great mathematicians in history and has sometimes been called the Newton of France. In the beginning of the 19th century he pictured the universe in a way that resembles the opinion of Einstein – stating that the universe is truly intelligible.

Closely related to this idea is the notion of causal determinism, demonstrated by a “demon”. Laplace did not use this word himself but referred to it as “une intelligence”.

> We may regard the present state of the universe as the effect of its past and the cause of its future. An intellect which at a certain moment would know all forces that set nature in motion, and all positions of all items of which nature is composed, if this intellect were also vast enough to submit these data to analysis, it would embrace in a single formula the movements of the greatest bodies of the universe and those of the tiniest atom; for such an intellect nothing would be uncertain and the future just like the past would be present before its eyes.

This idea closely resembles that of Stephen Hawking and many other scientists who are searching for the “formula of everything”, the ultimate goal of knowledge. It is also an echo of Descartes’ idea of nature as a big machine, possible to fully understand.

It is really ironic that Laplace calls his “demon” intelligent because it is difficult, if not impossible, to find anything more unintelligent. Laplace’s argument is self-contradictory.

In order to reach its goal – that nothing is uncertain – the intelligent creature has to leave the context of which it is part and take residence outside. It cannot possibly be a part of what it wants to know all about, because that would not permit the total view.

Thus the demon must be much worse off than being unintelligent. It must be dead (as dead as the necessary subjectivity it once had). If it is seeking omniscience, it must also be able to analyze and understand its own life from beginning to end, and that is impossible if it is alive. It has to wait for its life to terminate to give the final verdict. But at that point it is too late. Omniscience does not exist in the way the demon, or Laplace, thought. A living creature has to reach intelligibility in another way than obliterating itself and its subjectivity.

And what about quantum physics? Does it not disarm the demon as well, making scientific omniscience a dream of the past? Does it not say that it is impossible to simultaneously know "all forces that set nature in motion, and all positions of all items of which nature is composed"? If you describe the momentum of a particle accurately you cannot describe its position with the same accuracy. The greater the precision of the one, the lesser of the other. Nature does not seem eager to be caught in the net of intelligibility.

The world can be made intelligible by means of a "true model". This is the legacy of Bacon, Descartes and the Enlightenment. But can a "true model" really exist?

I think models can be true, but only if they are what they model. The only true model of the forest outside my window is exactly the same forest. If those are the conditions, true models could not be observed. How can we construe such an intricate, seething, manifold model? How can we even see it? The more complicated something is, the harder it is to construe a reliable model. Weather predictions are difficult to make, even if they have become vastly improved, because of the sheer complexity of weather phenomena.

These obstacles do not seem to deter humankind in the search for true models. This search is part of everyday life, in literature, art, music, theatre, film, religion, philosophy, science, politics and economy. The models and maps are getting increasingly more reliable. Or do they? Voices have been raised to the contrary. Models *explain* nothing.

The philosopher Friedrich Nietzsche dwells on the idea in *The Joyful Science.*

> 'Explanation' we call it, but 'description' is what distinguishes us in comparison to older forms of knowledge and science. We describe more accurately – explain we do no more than our predecessors. We have discovered a many-fold succession of events, where naive man and the researcher of older cultures only saw two phenomena, 'cause' and 'effect' as it is called. We have improved the picture of the course of events, but we have not come past the picture, or behind it. The succession of 'causes' is more complete in front of us in every single case, we draw our conclusions: this and this must first happen in order for this to follow – but that is not the same as understanding . . . How would it, by the way, be possible to explain! We solely operate with things that do not exist, with lines, surfaces, bodies, atoms, divisible

> times, divisible room – how would an explanation be at all possible, when we first turn everything into a picture, our own picture! It is enough to regard science as a most faithful way to make things human, we learn to describe ourselves more and more accurately by describing the succession of things.

Nietzsche is fully aware of the problems with a "true model". He simply says that we should not believe that we explain things when we merely describe and mimic them.

It seems to us, when we create models, that they are modelled on an independent reality, one that would be exactly the same without us. In so believing we omit the necessary subjectivity, do not see that *our participation is obligatory for the "independence" of reality.*

So, is the world fully intelligible? No, if by way of science, with the help of models. The model can never contain the whole truth because the model is a diversion of what it is modelling, otherwise it would not be a model. We should think of science as one of many languages, a way of dealing with nature and our surroundings, but not as a provider of truth. Only if the model-theory is transgressed, we can answer in the affirmative: *if there is a model which is an exact copy of what it is modelling, without ceasing to be a model.* I would say that a theory that contains its practice, and practice that contains its theory is such a model.

The requirements are harsh: if the model is to be the truth about us, it has to live. It cannot be a dead statements such as "this is humankind" or "this is the truth about our existence", because truth must deal with the totality of what it means to be human, and life is part of that.

Could there possibly be such a model showing us all that we want to know about ourselves and the world, a completely satisfying answer that will put an end to our alienation?

The goal is evident: if we want the truth about the world and ourselves we have to find a model that is the same as what it is modelling, a symbol that is what it symbolizes. Theory has to mirror practice in full (as a model) and practice theory in order for us to know the truth.

This is the only way for us to get to the truth. Theory and practice have to overlap completely. If they do not, theory will never explain practice and practice will not mirror theory. In that case theories will constantly need revision because they do not overlap with practice, and practice will sooner or later fail, because it lacks theory.

I think it is possible for all of us, and not a privilege of scientists, philosophers or founders of religions to find a model which is true, a map that is the forest. All of us have what it takes: language. The word. Language is both the lock and the key. It has a double identity. *It throws humankind out of Paradise and at the same time guarantees that we cannot leave it.*

In an earlier chapter I showed that in order for a self to become conscious of itself there has to be a deviation from what the self really is (even though it still is the same). I said that we are what we are even though we always, at the same time, are what we are not.

If this deviation is regarded as natural and everlasting, we will never reach the truth about our existence. The deviation has to be replaced by what we are to make it vanish. But does not language and our special human consciousness vanish at the same time? In an earlier chapter I described language as the cause of deviation making consciousness possible. You can say that this reverse process of our integration with what we are in a way extinguishes language. It turns it into a new kind of language: not one we have but a language that we are. And this is at the same time something that language always has been and cannot refrain from being.

Language has never been a mere instrument or a means of communication.

Let's use the word "human being" as an example. In what way can this word cease to be just theory and become a theory inseparable from our bodies? If it is just theory it will only be a model and we will never reach the truth about ourselves. It also has to be practice, and not *any* practice, but one that is the full rendering of the theory. This will be the true model.

For us to fully understand ourselves we have to be the word "human being". Our bodies have to be it in a way that there will be no difference between them and the word. We will be a living word, just as a true map of the forest would be the same, living forest. The word and the body will be two sides of the same thing. Through one we will see the other and vice versa. The word will laugh, talk, sing, eat, suffer and bleed. And do not say that this is impossible to achieve because it has been done, at least by one person in our history, a man who said he was the truth rather than using the misleading "this is the truth".

There is quite a step to be taken from "this is the truth" to "we are the truth". It involves a change of language and a change of world view as well as a change of society. "This is the truth" is a phrase used by people who regard themselves as secondary subjects, who think they are produced by some yet not fully understood power like Nature or God. "We are the truth" on the other hand are words used by people who understand that they are fully responsible for themselves as primary subjects. Primary subjects like Nature or God are there because we ourselves are secondary in unequal societies. The inequality gives us a thwarted view of the world and ourselves. So in order for us to become the words that we are, to be both practice and theory we have to move away from inequality towards equality.

How is this exchange of languages to be understood? It is the movement from a provisional, vacillating language into a permanent one. How often does not an expression like "human being" change its value or meaning during the course of a day or a week? We can try to stay with one particular meaning, but it still is not permanent. For this word to be everlasting it must comprise "what we are" and leave nothing behind. Whenever it is used, no matter the situation, it must remain the same. The persons, who use this word, also apply it to themselves. If their bodies, their lives, the idea of themselves, all they do and stand for is summed up by this word, there will be no need for a change. And moreover: we cannot choose "what we are". It chooses us. This is the only way that theory, the word, can become practice, the body. If the word is transformed in this way nothing can be deducted from it and nothing added. It will always remain the same, shining eternally. This is the same as saying that we are forever tied to ourselves, being ourselves in our "eternal validity".

How then will we experience ourselves? In what way will we have changed knowing ourselves in this infallible way? I think the change will be tremendous, unimaginable. When "what is not" has totally disappeared in favour of "what is", I believe that we no longer have it in our power to be devious, to pretend, to lie, to treat people differently, to manipulate, to admire or despise. Money, status, power, glory will be outshone by a brighter, steadier light: the full understanding of ourselves. We need a minimum of laws. Life no longer vacillates between happiness and unhappiness. Anxiety and fear are gone and so is shame and guilt. We are back in the paradise we never left and never will leave, locked to the word we are.

If a word in this way is complete, being the body, the ultimate goal is achieved. There should be nothing else to desire. You look straight into the furnace of you own creation, from this moment on at peace with yourself, totally approving of yourself. The double standard when people act against their own will that St Paul expressed in his letter to the Romans is no more. Such behavior evaporates in the light of truth: the subconscious is dispersed.

This road to ourselves is not to be mixed up with that of fanatics and fundamentalists. Their words are arbitrary, construed contrary to "what is". Fanatics "despairingly want to be themselves" as Kierkegaard put it, thereby always being counterproductive. The harder the fanatics work to reach their goal the farther away they get. This happens because they postulate a "what is" of their own liking, force a connection with something that is not there. They do not understand that they are already connected. Fanaticism and fundamentalism are two of inequality's many deceases, luring people into believing that they can be servants of truth. Such servants are harmful to themselves and humanity and will always be.

We have to live our words, rise with them in the morning and go to bed with them at night. If we say that all men are created equal, we have to live those words or they will mean nothing. To find out what they mean they have to be anchored in our actions and body. If we think that "humans are egoists" we have to live the words, live them if we presume that "humans are animals", live them if we are sure that there is "a seed of evil in us all".

Live all of it, do not just say it. If we succeed in making these ideas viable, we can trust them. They will never desert us. When it comes to my examples above I am convinced that we cannot really live them, any more than we can live "such is humankind" or "such is life". The harder we try the emptier the words will get. The end is nothingness.

Sometimes people say: "We must start with ourselves to change the world." This might be a reaction to all unsuccessful social experiments, like that of the Soviet Union. These experiments often start with a change in society and a hope for people to follow. But that hope is futile if people do not simultaneously start with themselves. The old language will be dragged along into the new society and nothing will really be gained. To achieve a real change the start must be dual. If you start with yourself this must also mean that you start with society. And if society is to begin, the start must involve its members. There will only be a real change if both language and society change, no matter which of the two takes the lead.

The relationship between humankind and nature, between a human being and herself, is much more complicated than our present models indicate. We have not just risen out of nature molded by a long and arduous process called evolution. There is more to our existence than that. We are, in a sense, by being conscious of the world, children of ourselves. Our roots with "what is" and with ourselves are not to be found in nature, but in the nature of language. The word is not just a word, but also our flesh and bone, something we are. It is reality.

It seems incredible that there is another life where we are our true masters, totally in peace with what we are. But to me that is the real life, hidden behind the veil of fundamental misconceptions, comparable to superstitions. If they are cleared away the sun will always shine to lighten our aspirations. I am quite sure that people who have accomplished this, or part of it, live a life without worries and anxiety, as certain of their identity as any animal, uninterested in questions like "who am I", "how was I created", "what is a human being".

These and similar questions are already answered by their predetermined life and words. I believe that no certitude is greater than theirs and no light more enduring.

Chapter 15
A conclusive assumption

It is evident that humankind causes its own sufferings. War, oppression, poverty, and most diseases are not caused by nature, although some natural disasters are. Is the explanation that we are all crooks and sinners? Many people, faced with our history, are prone to think so. But that is just a political statement used in order to justify their actions as is the talk of a "human nature". The true explanation is something different altogether. We suffer because we lack knowledge like the persons whose heads are fixed in Plato's cave. But our fetters are not visible, do not seem to exist, which enables us to talk about our "free will".

Our fixture is language, which makes us see ourselves and the world in a predetermined perspective rooted in the structure of our societies. Our beliefs, our science and culture make us dance to the tune of inequality even though our ears seem deaf to the sound. In order to create a new language with other predeterminations we have to move towards equality. All directions which tend to preserve inequality will only tear the body of humankind apart.

We cause our own suffering. The greatest of all sufferings is death. Maybe, if we are meticulous, we will see that death is something we inflict on ourselves. Death might very well be the price we pay for replacing true knowledge with beliefs and ephemeral ideas.

The last words of Socrates in Plato's Republic echo what I have just implied. We should not live in the light of death as the Buddhists recommend but in the light of eternal life:

> This at any rate is my advice, that we should believe the soul to be immortal, capable of enduring all evil and all good, and always keep our feet on the upward way and pursue justice and wisdom. So we shall be at peace with God and with ourselves, both in our life here and when, like the victors in the games collecting their prizes, we receive our reward, and both in this life and in the thousand-year journey which I have described all will be well with us.

Literature

This book refers to texts in both English and Swedish. A Swedish translation of a book or text in English has the title in Swedish first in the list followed by the original English title. A book or text by a Swedish author has the title in Swedish first and, as a service to the reader, the title translated into English. If any of these Swedish books are translated and published – very few of them are – the published English title might be different from the one I have chosen. But this problem should be easy to overcome, if anyone wants to find the English edition. So if you see, for example: *Dawkins, Richard. 2006. Illusionen om Gud. The God Delusion. Leopard förlag*, you know that I have read the Swedish edition and you also know the English original title and the Swedish publisher. In one or two instances, when I did not find the English original, I have translated the Swedish text of a quoted passage back to English.

Bibliography

Alcock, John. 2001. The Triumph of Sociobiology. Oxford University Press.

Andersson, Sten. 2012. Filosofen som inte ville tala. The Philosopher who would not speak. Norstedts.

Bacon, Francis. 1960. The New Organon, Bobbs-Merrill, Indianapolis.

Bacon, Francis. 1967. Encyclopedia of Philosophy, MacMillan Free Press

Bacon, Francis. 1984. Essays. (Essäer). Daidalos

Behe, Michael, 1996. Darwin's Black Box, The Free Press

Berg, Lasse. 1994. Världens ålder. The Age of the World. Rabén Prisma

Berg, Lasse. 2005. Gryning över Kalahari. Dawn over Kalahari. Ordfront förlag

Blackmore, Susan. 1999. The Meme Machine. Oxford University Press

Blackmore, Susan. 2005. Consciousness, a short Introducation. Oxford University Press

Bourdieu, Pierre. 1993 Kultursociologiska texter. Culture and Sociology. Salamander

Bourdieu, Pierre. 2000. Konstens regler. The Rules of Art. Brutus Östlings bokförlag

Bourdieu, Pierre. 1998. Om televisionen. About Television. Brutus Östlings bokförlag

Bourdieu, Pierre. 1999. Den manliga dominansen. The Male Dominance. Daidalos

Brown, Donald E. 1991. Human Universals. http://condor.depaul.edu/mfiddler/hyphen/humunivers.htm

Budiansky, Steve. 1998. If a Lion could talk. Orion Books Ltd

Chomsky, Noam. 1969. Language and Mind. Människan och språket. Norstedts

Coetzee, J M. 2003 Nobel Lecture.

Cosmides, Leda. John Tooby. 1997. Evolutionary psychology: A primer http://www.cep.ucsb.edu/primer.html

Dammann, Erik. 1980. Revolution i överflödssamhället. Revolution in an affluent Society. Askild & Kärnekull

Dammann, Erik. 1978. Tio steg mot framtiden. Ten Steps into the Future. Askild & Kärnekull

Danielsson, Ulf. 2008. Den bästa av världar. The best of Worlds. Albert Bonniers förlag

Danielsson, Ulf. 2003. Stjärnor och äpplen som faller. Stars and Apples that fall. Albert Bonniers förlag

Darwin, Charles. 1871. The Descent of Man. Prometheus Books

Davies, Paul. 1999. The fifth Miracle. Penguin Books

Dawkins, Richard. 2009. Så gick det till. The Greatest Show on Earth. Leopard förlag

Dawkins, Richard. 2006. Illusionen om Gud. The God Delusion. Leopard förlag

Dawkins, Richard. 1976. The selfish Gene. Den själviska genen. Tidens förlag

Dawkins, Richard, 2011. The Magic of Reality, Transworld Publishers, Black Swan

Dawkins, Richard. 1995. Livets flod. River out of Eden. Natur & Kultur

Dennett, Daniel C. 1996. Darwin´s dangerous idea. Penguin

Deleuze, Gilles https://archive.org/details/DesertIslandsAndOtherTexts

Dunbar, Robin. 2004. Historien om människan. The human Story. Dualis

Dyson, Freeman. 1988. Infinite in all Directions. Harper & Row

Edberg, Rolf, 1972. Vid trädets fot. At the Foot of the Tree. Norstedts

Edberg, Rolf. 1966. Spillran av ett moln. The Remains of a Cloud. Norstedts

Edberg, Rolf. Jablokow, Alexej. 1988. Söndag är för sent. Sunday is too late. Norstedts

Edberg, Rolf. 1974. Ett hus i kosmos. A House in Cosmos. Esselte Studium

Edberg, Rolf. 1987. Årsbarn med plejaderna. Born the same Year as the Pleiads. Bra Böcker

Edman, Stefan. 2003. Jorden har feber. The Earth has a Temperature. Bokförlaget Atlas

Edman, Stefan. 1981. Jord att älska. Earth to love. Gummessons

Edman, Stefan. 1984. Jordens sang. The Song of the Earth. Norstedts

Edman, Stefan. 2006. Förundra, tankar om vår stund på jorden. Wonder, Thoughts about our Time on Earth. Cordia/Verbum

Eriksen, Thomas Hylland. 1996. Historia, Myt och Identitet. History, Myth and Identity. Bonniers

Ernberg, Ingemar and others. 2010. Vad är liv, i kosmos, i cellen, i människan? What is Life, in Cosmos, in the Cell, in Humankind? (Karolinska Institutet. University Press

Fish, Stanley, 2001. The trouble with Principle. Harvard University Press

Fromm, Eric. 1976. To have or to be. Abacus

Garff, Joakim. 2002. Sören Aabye Kierkegaard, en biografi. Nya Doxa

Gordimer, Nadine. 1991. Nobel Lecture http://www.nobelprize.org/nobel_prizes/literature/laureates/

Gould Stephen Jay. 1983. Den felmätta människan. The Mismeasure of Man. Bokförlaget Alba

Grass, Günther. 1999. Nobel Lecture. http://www.nobelprize.org/nobel_prizes/literature/laureates/

Greene, Brian. 1999. Ett utsökt universum. The elegant Universe. Norstedts

Gustafsson, Martin. http://mfaa.abo.fi/?article=vad-menar-vi-med-intelligens (what do we mean by intelligence?)

Haugeland, John. 1985. Artificial Intelligence – The Very Idea. MIT Press

Hawking, Stephen. Mlodinow Leonard. 2010. Den stora planen. The grand Design. Norstedts

Hawking, Stephen. 1988. Kosmos, en kort historik. A brief History of Time. Prisma

Heany, Seamus. 1995. Nobel Lecture. http://www.nobelprize.org/nobel_prizes/literature/laureates/

Heberlein, Ann. 2011. A small Book of Evil. En liten bok om ondska. Bonniers

Horkheimer, Max, Adorno, Theodor. 1947. Upplysningens dialektik. Dialectics of Enlightenment. Röda bokförlaget

Hubendick, Bengt. 1974. A small Book about Man on Earth. En liten bok om människan på jorden. Zindermans

Hubendick, Bengt. 1985. Människoekologi. Human Ecology. Zindermans

Hume, David. 1748. An Enquiry concerning Human Understanding, Dover Publications inc.

Israel, Joachim och von Heijne, Gunnar. 1985. Skapelsens herre eller genernas slav? The Lord of Creation or the Slave of Genes? Ordfront förlag

Jackendoff, Ray, 1995. Patterns in the Mind, Language and Human Nature, BasicBooks, Perseus Books

Jackson, Tim. 2009. Välfärd utan tillväxt. Prosperity without Growth. Ordfront förlag

Jersild, P C. 1997. Darwins ofullbordade. Darwin's unfinished. Bonnier Alba Essä

Jonstad, David. 2011. Kollaps. Collapse. Ordfront

Judt, Tony. 2010. Illa far landet. Ill fares the Land. Karneval förlag.

Jung, C G. 1989. Mitt liv. Memories, Dreams, Reflections. Wahlström & Widstrand

Jung, C G.1977. Om psykisk energi, drömmar och arketyper. About psychic Energy, Dreams and Archetypes. Wahlström & Widstrand.

Kant, Immanuel. 1989. Vad är Upplysning? What is Enlightenment? Symposion

Kant, Immanuel. 1793. 1960. Religion within the Limits of Reason, Harper & Brothers

Kauffman, Stuart, 1996. At Home in the Universe. Penguin

Kemp, Peter. 1979. Språk och existens. Language and Existence. Liber förlag

Kemp, Peter, 1972. Sprogets dimensioner. The Dimensions of Language. Berlingske

Kértesz, Imre. 2002. Nobel Lecture. http://www.nobelprize.org/nobel_prizes/literature/laureates/

Kierkegaard, Sören. 1844. 1996. Begreppet ångest.The Concept of Anxiety. Nimrod

Kierkegaard, Sören. 1846. 1999. Avslutande ovetenskaplig efterskrift I, II. Concluding unscientific Postscipt I, II. Nimrod

Kierkegaard, Sören. 1849. 1996. Sjukdomen till döden. Illness unto Death. Nimrod

Krauss, Lawrence. 2012. A Universe from Nothing, Simon & Schuster Ltd

Leaky, Richard. 1981. Making of Mankind. Hur människan blev till. Natur & Kultur

le Clezio, J M G. 2008. Nobel Lecture. http://www.nobelprize.org/nobel_prizes/literature/laureates/

Lessing, Doris. 2007. Nobel Lecture. http://www.nobelprize.org/nobel_prizes/literature/laureates/

Lindborg, Rolf, 1968. René Descartes. Natur & Kultur

Lindley, David. 1994. The End of Physics. BasicBooks

Lindström, Jonathan. 2013. Medvetandets gåta, om hjärnan, själen och konsten att kila runt hörnet. The Enigma of Consciousness, about the brain, the soul and the art of dying. Norstedts

Lipton, Bruce. 2007. The Biology of Belief. Hay House Inc.

Llosa, Mario Vargas. 2010. Nobel Lecture. http://www.nobelprize.org/nobel_prizes/literature/laureates/

Maclagan, David. 1977. Creation Myths. Thames and Hudson

Magee, Bryan, 1974. Popper. Fontana/Collins

Malcolm, Norman. 1996. Minnen av Wittgenstein. Ludwig Wittgenstein, a Memoir. Thales

Malmberg, Bertil. 1970. Språket och människan. Language and Humankind. Aldus/Bonniers

Milton, John. 1667. 2005. Paradise Lost. Dover Editions.

Monk, Ray.1992. Ludwig Wittgenstein. Daidalos

Müller, Hertha. 2009. Nobel Lecture. http://www.nobelprize.org/nobel_prizes/literature/laureates/

Nietzsche, Friedrich, 1882. 2011. Den glada vetenskapen. Joyful Science. Bokförlaget Korpen

Nörretrenders, Tor. 1999. Märk världen. Mark the World. Bonnier Alba

O'Hear, Anthony, 1997.Beyond Evolution: Human nature and the limits of evolutionary explanations, Oxford University Press

Olson, Lars, Josephson, Anna and others. 2012. Hjärnan. The Brain. Karolinska Institutet

Oz, Amos. 2009. How to Cure a Fanatic. Hur man botar en fanatiker. Wahlström & Widstrand

Pamuk, Orhan. 2006. Nobel Lecture http://www.nobelprize.org/nobel_prizes/literature/laureates/

Paz, Octavio. 1990. Nobel Lecture. http://www.nobelprize.org/nobel_prizes/literature/laureates/

Pinker, Steven. 1994. The Language Instinct, Penguin Books

Pinker, Steven. 2002. The Blank Slate, Penguin Books

Pinter, Harold. 2005. Nobel Lecture. http://www.nobelprize.org/nobel_prizes/literature/laureates/

Plato. 2007. The Republic. Penguin Classics

Platon. 2000. 2001. Skrifter, Bok I, Bok II. red Jan Stolpe. Dialogues. Atlantis

Porter, Roy. 1997. Upplysningen. Enlightenment. Daidalos

Prigogine, Ilya. 1997. The End of Certainty. The Free Press

Putnam, Hilary. 1991. Representation and Reality. MIT Press

Quine, William Van Orman. 1992. Strävan efter sanning. Pursuit of Truth. Nya Doxa

Ricoeur, Paul. 1991. Från text till handling. From Text to Action. Symposion

Rockström, Johan, Klum, Mattias. 2012. Vår tid på jorden. Our Time on Earth. Bokförlaget Langenskiöld

Rogers, Lesley, 2002. Sexing the Brain. Weidenfeld & Nicolson

Rousseau, Jean-Jacques. 1958. The Social Contract. Discourses. Dent & Sons Ltd.

Sartre, Jean-Paul. 1986. Existentialismen är en humanism. L'existentialisme est une humanism. Askelin & Hägglund

Sartre, Jean-Paul. 1960. 1984. Till frågan om metoden. On the Question of Method. Arkiv förlag

Sartre, Jean-Paul. 1957. The Transcendence of the Ego. The Noonday Press

Sartre, Jean-Paul. 1943. 1983. Varat och Intet. Being and Nothingness. Bokförlaget Korpen

Satel, Sally and Scott O Lilienfeld, Brainwashed. 2013. The Seductive Appeal of Mindless Neuroscience. Basic Books

Sjölander, Sverre. 2002. Naturens budbärare. The Messengers of Nature. Nya Doxa

Spinoza, Baruch. 1955. Works of Spinoza I och II. Dover Publications Inc

Stevens, Anthony. 1992. Jung, om hans liv och verk. Jung, on his life and work. Gedins

Sörlin, Sverker, Öckerman, Anders. 2000. Jorden en ö, en global miljöhistoria. The Earth an Island, a Global Environmental History. Natur och Kultur

Sörlin, Sverker. 1992. Human Ekologi. Human Ecology. Carlssons

Tattersall, Ian. 1999. Becoming Human. Harcourt Brace & Company

Ulfstrand, Staffan. 2011. Darwins idé. Darwin's Idea. Symposion

Uddenberg, Nils. 1993. An Animal among Others. Ett djur bland alla andra. Nya Doxa

Uddenberg, Nils. 1998. Arvsdygden. The Heritage of Virtue. Natur och Kultur

Uddenberg, Nils. 2003. Idéer om livet, I och II. Ideas about Life I and II. Natur och Kultur

Warnock, Mary. 1966. Sartres filosofi. The Philosophy of Sartre. Rabén & Sjögren

Williams, Bernard. 2004. Truth and Truthfulness. Princeton University Press

Wijkman, Anders, Rockström, Johan. 2011. The great Denial.Den stora förnekelsen. Medströms bokförlag

Wittgenstein, Ludwig. 1967. Philosophical Investigations. Blackwell Publishers.

Wittgenstein, Ludwig. 1992. Filosofiska undersökningar. Philosophical Investigations. Thales

Wittgenstein, Ludwig. 1972. On Certainty. Harper Torchbooks

Wittgenstein, Ludwig. 2000. Filosofi. Philosophy. Glänta Produktion

Wittgenstein, Ludwig. 1995. Zettel. Thales

Wittgenstein, Ludwig. Lectures & Conversations. (No print year). University of California Press

Wittgenstein, Ludwig. 1999. Blå boken och Bruna boken. Blue and Brown Book. Thales

von Wright, Georg Henrik. 1986. Science and Reason. Vetenskapen och förnuftet. Bonniers

von Wright, Georg Henrik. 1978. Humanismen som livshållning. Humanism as an Attitude to Life. Rabén & Sjögren

von Wright, Georg Henrik. 1993. The Progress Myth. Myten om framsteget. Bonniers

von Wright, Georg Henrik. 1994. Att förstå sin samtid. Understanding the Contemporary. Bonniers

Worster, Donald. 1977. Nature's Economy, Cambridge University Press

Worster, Donald. 1994. The Wealth of Nature. Oxford University Press

Xingjian, Gao. 2000. Nobel Lecture. http://www.nobelprize.org/nobel_prizes/literature/laureates/

Initials used as references in the text

AJ — Alexej Jablokow. Born in 1933. Russian biologist and founder of the Greenpeace branch in Russia. Chairman of Greenpeace in Russia until 1991.

RE — Rolf Edberg. 1912-1997. Author of a large number of books about the environment. Ambassador to Norway and Governor of Värmland.

AW — Anders Wijkman. Born in 1944. Served several periods as member of the Swedish Parliament. Author of books about the environment.

JR — Johan Rockström. Born 1965. Professor of environmental science at the University of Stockholm. Author of books on ecology and related subjects.

MK — Mattias Klum. Born 1968. World known National Geographic photographer, author of books about the environment.

SE — Stefan Edman. Born 1946. Advisor to the Swedish Government on environmental issues. Writer of many books on the environment.

LB — Lasse Berg. Born 1943. Writer of a great number of books on the origin of humankind.

UD — Ulf Danielsson. Born 1964. Professor of theoretical physics at the University of Uppsala. Has popularized science in three books.

SU — Staffan Ulfstrand. Born 1933. Professor of ecological zoology at the University of Uppsala. Has written several books on ecology and zoology.

AH — Ann Heberlein. Born 1970. Doctor of theology at the University of Lund.

IT — Ian Tattersall. Born 1945. Paleoanthropologist and a curator emeritus with the American Museum of Natural History in New York.

NU Nils Uddenberg. Born in 1938. Associate Professor of psychiatry at the University of Lund.

JL Jonathan Lindström. Born in 1962. Archaeologist and author of science books about conscience.

Index

E

F

G

H

I

J

K

L

M

N

O

P

R

S

T

U

V

www.ingramcontent.com/pod-product-compliance
Lightning Source LLC
LaVergne TN
LVHW022043200926
848481LV00003B/772
9781622731435